Vapor Intrusion Simulations and Risk Assessments

Yijun Yao · Qiang Chen

Vapor Intrusion Simulations and Risk Assessments

Springer

Yijun Yao
Institute of Soil Science
Chinese Academy of Sciences
Nanjing, China

Qiang Chen
State Environmental Protection Key Laboratory of Soil Environmental Management and Pollution Control
Nanjing Institute of Environmental Sciences of the Ministry of Ecology and Environment
Nanjing, China

ISBN 978-981-19-2699-0 ISBN 978-981-19-2700-3 (eBook)
https://doi.org/10.1007/978-981-19-2700-3

This Springer imprint is published by the registered company Springer Nature Singapore Pte Ltd.
The registered company address is: 152 Beach Road, #21-01/04 Gateway East, Singapore 189721, Singapore

Acknowledgment

This work is supported by the National Key Research and Development Program of China (nos. 2021YFC1809103 and 2020YFC1807002) and the National Natural Science Foundation of China (nos. 42107066 and 41771494).

Contents

About the Authors

Yijun Yao is a professor at the Institute of Soil Science, Chinese Academy of Sciences. His major research interest involves site investigation, risk assessments, and management of soil pollution. He is the author of several risk assessment tools, including the CVI2D and PVI2D toolkits, and Yao's spreadsheet for risk assessments for contaminated sites. He graduated from Brown University with a Ph.D. degree in Environmental Engineering in 2012.

Qiang Chen is an assistant researcher in Nanjing Institute of Environmental Sciences of the Ministry of Ecology and Environment, with a research focus on investigation and risk assessment of contaminated sites, environmental policy and decision-making, and software development. He has a master's degree in Environmental Engineering from Hohai University and is now pursuing a doctoral degree under the supervision of Prof. Yijun Yao.

Screening Models of Vapor Intrusion 1

1.1 Introduction

Vapor intrusion (VI) models simulate the transport of contaminant soil gas from subsurface sources into the buildings of concern at contaminated sites, concentrating on the prediction of the contaminant concentration attenuation (i.e., concentration reductions) relative to a subsurface concentration during the soil gas transport process and the migration into a building, as shown in Fig. 1.1 [2]. There are two kinds of VI models. One refers to simple risk screening tools, usually one-dimensional (1D) analytical models [3]. The other includes the numerical models, mostly multidimensional and used to study the influences of environmental factors in specific VI scenarios. In the first kind, the prediction may not be so accurate due to the over-simplification of the complicated reality. In contrast, numerical models can simulate more sophisticated cases but require significant computational effort and specialized knowledge, limiting their use by ordinary site investigators. Nonetheless, the differences are not absolute, and predictions by screening models can still be reliable if most essential factors are included in specific VI scenarios. On the other hand, numerical models have become more accessible to ordinary investigators due to the more user-friendly interface of modeling software and the advanced computational capacity of current personal computers. This chapter aims to overview the first kind of VI models with analytical mathematical solutions or VI screening models.

In the classic review of VI models by Yao et al. [2], screening models are considered partially based on the previous work on pesticide transport in soil [4–9] and radon vapor intrusion [10–13]. Usually, the soil gas transport part of VI screening models can be viewed as built upon the previous pesticide models [4–7], while the process of entry into the building benefited greatly from earlier studies of radon VI [10–13]. Though the subject has changed from radon to volatile organic compounds (VOCs) such as chlorinated solvents and petroleum products, the general governing equation still works.

Y. Yao, Q. Chen, *Vapor Intrusion Simulations and Risk Assessments*,
https://doi.org/10.1007/978-981-19-2700-3_1

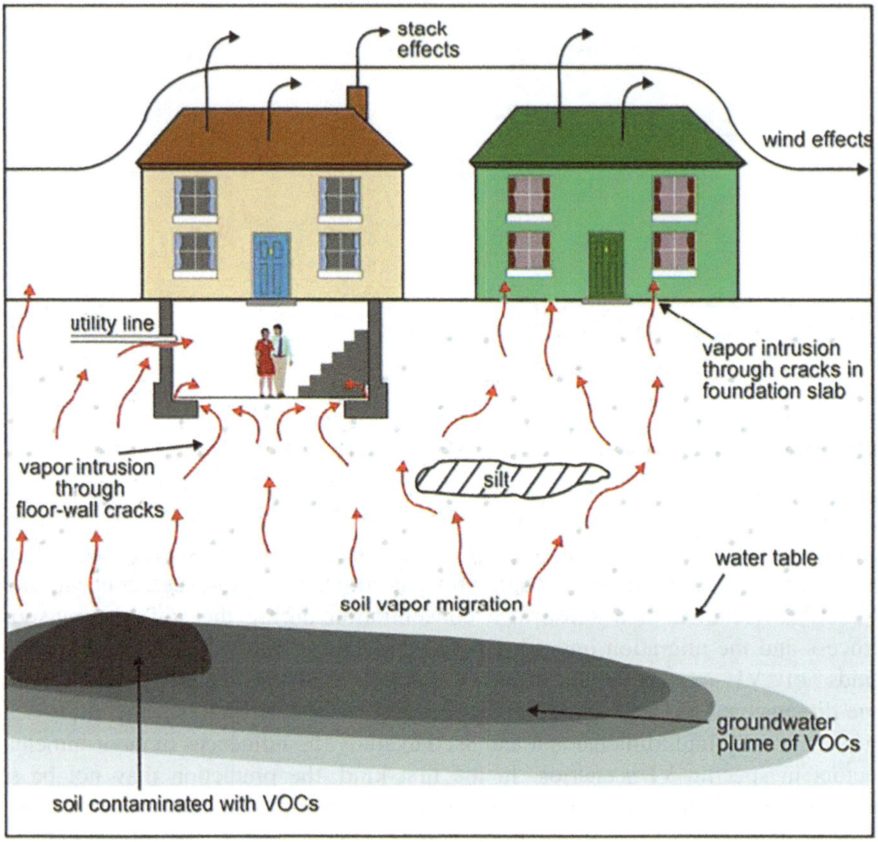

Fig. 1.1 Vapor intrusion [1]

The earliest pesticide models were developed in the late 1960s [14, 15] and have been used to study the fate and transport of pesticides after the injection into the soil. The processes simulated by pesticide models usually include leaching, volatilization, biodegradation, and so on. The pesticide models also consider factors like surface runoff, evapotranspiration, absorption, or drainage in site-specific cases. Obviously, only a part of these elements are relevant to VI and thus included in VI simulations.

Radon intrusion models were first introduced in the 1970s [10, 12, 13, 16, 17]. The radon soil gas is believed to enter the indoor space through the foundation cracks or permeable walls, mainly through advection caused by indoor-outdoor pressure differences. Such a pathway is also applied in VI involving VOCs, the source of which however is located separately from the target building instead of uniformly distributed in the sub-foundation soil as the radon source. This difference is why molecular diffusion due to concentration gradients plays a more important role in VI involving VOCs than radon intrusion.

The first VI screening model was the classic Johnson-Ettinger (J-E) model introduced in 1991 [18], which is also the most popular one. A typical VI screening model commonly includes the complete process from the vapor source (e.g., the contaminated groundwater) to the building of concern, including soil gas transport, entry into the building, and the calculation of indoor air concentration, as concluded by Yao et al. [2]. For the soil gas transport part, the major differences among screening models are the simplifications of the same general governing equation of convection and diffusion [2]. Then, the screening models employ different assumptions of the entry route through the building foundation, usually involving a dirty floor, a permeable slab, or a slab crack. At last, with the calculated contaminant entry rate into the building, the indoor air concentration can be predicted by assuming the indoor space as a continuous-stirred tank. In other words, temporal or spatial variations in contaminant indoor air concentration were not considered in the calculation of indoor concentration in a steady-state. In some cases, an empirical contaminant concentration attenuation factor was used as an alternative to estimating the indoor air concentration with the subslab soil gas concentration.

Contaminant vapor transport in the soil involves advection, diffusion, and other mass change terms, such as biodegradation and absorption/desorption, the latter of which only plays a role in transient cases [2]. Currently, it is commonly accepted that the dominant mechanism of dominant soil gas transport in the vadose zone is diffusion [19, 20], as confirmed by comprehensive numerical simulations of three-dimensional (3D) VI scenarios [21, 22]. Compared to diffusion, the soil gas advection only affects a small region in the sub-foundation, where the soil permeability is generally higher than that in the deeper soil. The advective soil gas flow is driven by pressure gradients, usually induced by the indoor-outdoor pressure difference [18] or indoor pressure fluctuations [23]. Many concepts and processes of radon entry into the building were reintroduced in the development of VI models [11–13, 24]. For instance, the Nazaroff equation [25], once developed to estimate the volumetric flow rate of radon soil gas into a pipe, was employed by screening models, such as the J-E model [18], to calculate the soil gas entry rate of VOCs into the building of concern.

At many VI sites, the contaminant vapor source is polluted groundwater. In other cases, the source could be contamination in the form of free nonaqueous phase liquid (NAPL) in the vadose zone [2]. For groundwater source, it is common to calculate the corresponding vapor concentration based on measured groundwater concentration and Henry's law [26, 27]. In the latter scenario, usually involving a much higher contaminant soil vapor concentration than the former, the direct measurement of soil gas is more recommended for a developed contaminant site [26]. The screening models usually characterize the vapor source, regardless of source types, with only a source vapor concentration and depth, except that sometimes an additional dilution factor of one-tenth was employed to include the attenuation through the capillary fringe in groundwater source cases [24].

To help understand the critical factors and assumptions employed, the current VI screening models are reviewed in an order based upon the consideration of sub-modules, as proposed by Yao et al. [2] First are presented the "soil gas transport modules" (including those involving no biodegradation and those that do). Next are the "entry pathway processes," which define how the contaminant soil gas enters the building with its mass entry rate. After that, we consider the "indoor air concentration calculation," which relates the contaminant mass entry rate to its indoor concentration. At last, some discussions of VI screening tools are provided.

1.2 The General Equation Governing Soil Gas Transport

Different from radon intrusion, it is essential to consider soil gas transport in the vadose zone for dislocated vapor sources, as in VI involving VOCs. In most screening models, diffusion dominates the soil gas transport process before the contaminant reaches the foundation slab. In other models, advection, biodegradation, and absorption are also considered to influence soil gas transport. The governing equation of advective soil gas flow is Darcy's Law, with magnitude and direction determined by the pressure gradient field. In most cases of VI involving VOCs, the pressure gradients are insignificant with respect to atmospheric pressure, and the assumption of incompressible soil gas is employed [2]

The general equation of Darcy's Law is:

$$\frac{\partial p}{\partial t} - \frac{p_{\mathrm{atm}} k_{\mathrm{g}}}{\phi_{\mathrm{g}} \mu_{\mathrm{g}}} \nabla \cdot (\nabla p) = 0 \tag{1.1}$$

$$q_{\mathrm{g}} = \frac{k_{\mathrm{g}}}{\mu_{\mathrm{g}}} \nabla p \tag{1.2}$$

where p is the soil gas pressure [M/L/T^2], t is time [T], p_{atm} is atmospheric pressure [M/L/T^2], k_{g} is the soil permeability [L^2], ϕ_{g} is the air-filled porosity [$\mathrm{L^3_{gas}/L^3_{soil}}$], μ_{g} is the soil gas viscosity [M/L/T], and q_{g} is the soil gas flux [$\mathrm{L^3_{gas}/L^2_{soil}}$ /T]. The effects of density and gravity are not considered in these equations [2].

For contaminant mass transport, the governing advection-diffusion equation [21] can be written as (1.3):

$$\phi_{\mathrm{g,w,s}} \frac{\partial c_{\mathrm{ig}}}{\partial t} = -\nabla \cdot \left(q_{\mathrm{g}} c_{\mathrm{ig}}\right) - \nabla \cdot \left(\frac{c_{\mathrm{ig}}}{H_{\mathrm{i}}} q_{\mathrm{w}}\right) + \nabla \cdot \left(D_{\mathrm{i}} \nabla c_{\mathrm{ig}}\right) - R_{\mathrm{i}} \tag{1.3}$$

where

$$\phi_{\mathrm{g,w,s}} = \phi_{\mathrm{g}} + \frac{\phi_{\mathrm{w}}}{H_{\mathrm{i}}} + \frac{k_{\mathrm{oc,i}} f_{\mathrm{oc}} \rho_{\mathrm{b}}}{H_{\mathrm{i}}} \tag{1.4}$$

In Eq. (1.3), $\phi_{\mathrm{g,w,s}} \frac{\partial c_{\mathrm{ig}}}{\partial t}$ is the transient term of contaminant mass in total three phases as defined in Eq. (1.4); $-\nabla \cdot \left(q_{\mathrm{g}} c_{\mathrm{ig}}\right) - \nabla \cdot \left(\frac{c_{\mathrm{ig}}}{H_{\mathrm{i}}} q_{\mathrm{w}}\right)$ refers to the advection term of contaminant movement in the form of gas and, if relevant, water phases; $\nabla \cdot (D_{\mathrm{i}} \nabla c_{\mathrm{ig}})$ is the term of molecular diffusion; ϕ_{w} is the water-filled porosity [$\mathrm{L^3_{water}/L^3_{soil}}$]; H_{i} is the Henry's Law constant of contaminant i [($\mathrm{M_i/L^3_{gas}}$)/($\mathrm{M_i}$/

Table 1.1 The summary of analytical models involving no biodegradation

Main governing equation	Examples of model use
Steady-state advection $0 = -\nabla \cdot (qc_{ig})$ (1.6)	Little et al. (#3) [30]
Steady-state diffusion $0 = D_i\nabla^2 c_{ig}$ (1.7)	OCHCA [31], J-E model [18], Park [32], CSOIL [33], Atteia and Hohener [29], Yao et al. [22, 34, 35][a], Shen and Suuberg 2014 [36][a]; Wrighter and Howell [37], Lowell and Eklund [38][a] McHugh et al. [39][a] CVI2D [40][a] Feng et al. 2020 [41]
Transient diffusion $\phi_{g,w,s}\frac{\partial c_{ig}}{\partial t} = \nabla^2 c_{ig}$ (1.8)	Little et al. (#1, 2 model) [30] Symms et al. [42].
Steady-state uniform advection and diffusion $q \nabla c_{ig} = \nabla \cdot (D_i \nabla c_{ig})$ (1.9)	Murphy and Chan [43], VOLASOIL [44, 45], CSOIL 2000 [46]

[a]Means 2D models, and models shown without superscript symbols are all 1D models

$L^3_{water})$],]; $k_{oc,\ i}$ is the sorption coefficient of contaminant i to organic carbon in the soil [$(M_i/M_{oc})/(M_i/L^3_{water})$]; f_{oc} is the mass fraction of organic carbon in the soil [M_{oc}/M_{soil}], ρ_b is the soil bulk density [M_{soil}/L^3_{soil}], c_{ig} is the concentration of contaminant i in the gas phase [M_i/L^3_{gas}], q_w is the water flux [$L^3_{water}/L^2/T$], D_i is effective diffusion coefficient of contaminant i [L^2/T], R_i is reaction rate of the contaminant i [$M_i/L^3_{soil}/T$], and $\phi_{g,\ w,\ s}$ is the effective transport porosity [L^3_{air}/L^3_{soil}], as defined in Eq. (1.4). Equations (1.3) and (1.4) apply not only for contaminant but also for oxygen transport in soil, as in some cases, the latter should be considered for simulating oxygen-limited biodegradation of petroleum products [2].

The effective diffusion coefficient of contaminant i, D_i can be calculated with the Millington and Quirk equation [28] (1.5):

$$D_i = D_i^g \frac{\phi_g^{\frac{10}{3}}}{\phi_t^2} + D_i^w \frac{\phi_w^{\frac{10}{3}}}{\phi_t^2} \quad (1.5)$$

where D_i^g is the contaminant diffusivity in air [L^2/T], ϕ_t is the total soil porosity [L^3_{pores}/L^3_{soil}], and D_i^w is the contaminant diffusivity in water [L^2/T] [2]. It should be noted that the mechanical dispersion is not considered in Eq. (1.5) [29]. In screening models, the common governing Eq. (1.3) for soil gas transport are usually simplified in different forms for specific VI scenarios, as shown in Tables 1.1 and 1.2 for non-biodegradation and biodegradation cases, respectively. Thus, the discussions in

Table 1.2 The summary of contaminant soil gas transport models, including biodegradation

Governing equation	Models
Steady-state diffusion and biodegradation (1 species) $0 = \nabla \cdot (D_i \nabla c_{ig}) - R_i$ (1.20)	Ririe and Sweeney [47], R-UNSAT [48], Johnson et al. [20], Davis et al. [49]
Transient diffusion and biodegradation (1 species) $\phi_{g,w,s} \frac{\partial c_{ig}}{\partial t} = \nabla \cdot \left(D_i \nabla c_{ig}\right) - R_i$ (1.21)	Jeng et al. [50], VIM [51], Sanders and Stern (#1) [52]
Constant advection and transient diffusion and biodegradation (1 species) $\phi_{g,w,s} \frac{\partial c_{ig}}{\partial t} = -q \nabla c_{ig} + \nabla \cdot \left(D_i \nabla c_{ig}\right) - R_i$ (1.22)	Ünlü et al.[53], Jury et al. [4–7], Van Genuchten et al. [54–56], Sanders and Stern (#2) [52], Anderson et al. [57], T & R model [58], Lin and Hildemann [59]
Steady-state diffusion and biodegradation (2 species) $0 = \nabla \cdot (D_i \nabla c_{ig}) - R_i$ (1.23)	Ostendorf and Kampbell [60, 61], Roggemans et al. [61, 62], Devaull [63], Parker [64], Verginelli and Baciocchi [65], PVI2D [66][a]
Steady-state diffusion and uniform advection and biodegradation (2 species) $0 = -q \nabla c_{ig} + \nabla \cdot (D_i \nabla c_{ig}) - R_i$ (1.24)	MVI [67]

Adapted with permission from [2]. Copyright 2013 American Chemical Society

[a]Means 2D model; models shown without superscript symbols are all 1D, steady-state, and analytical models

this chapter will be divided into two parts, depending upon whether or not biodegradation is explicitly included in the governing equation.

1.2.1 Soil Gas Transport Involving No Biodegradation

Most VI screening models were developed to simulate only steady-state cases and are based on analytical **solutions** of the one-dimensional (1D) transport equation. Table 1.1 summarizes the most available VI screening models involving no biodegradation and the corresponding governing equations. In 1D models, the "del" notation in the left column of Table 1.1 simply refers to "d/dz," i.e., an ordinary 1D derivative in the vertical direction. In the 2D models, the "del" notation represents partial derivatives in vertical and horizontal directions [2].

1.2.1.1 1D Screening Models Involving No Biodegradation

It is commonly believed that diffusion dominates soil gas transport of non-biodegradable contaminants with moderate or low soil permeabilities [18, 21, 22, 38]. The advective soil gas flow caused by an indoor-outdoor pressure difference or temporal variation in indoor pressure only plays a role in a small high-permeability region near the foundation. This is the reason that advection is not included in most 1D models. As described below, Yao et al. [2] summarized VI screening models based on governing equations for non-biodegradable contaminants.

Both the Orange County Health Care Agency (OCHCA) VI model [31] and Wrighter and Howell's study in 2004 [37] employed a simplified Eq. (1.3) to simulate a steady-state diffusive soil gas transport of contaminant from a subsurface source. In the OCHCA VI model [31], the upward diffusion rate is calculated by assuming a negligible subslab contaminant concentration. In the OCHCA model, the contaminant indoor air concentration was estimated by employing an empirical attenuation factor for the diffusion rate from the subsurface source due to the presence of the foundation slab.

Like the OCHCA VI model [31], the classic J-E model [18] is also based on a 1D diffusion-based equation but does not assume a negligible subslab concentration. Moreover, the building footprint size in the former was replaced with the building foundation area in contact with soil as the source size in the latter. The J-E model has been included in the EPA spreadsheet for modeling subsurface vapor intrusion [68], and commercial risk assessments tools, such as BP's Risk-Integrated Software for Cleanups (RISC) [69] and RBCA toolkit [70]. An equilibrium partitioning was also introduced in the J-E model by Park [32] for petroleum chemicals.

Little et al. in 1991 [30] developed several models based on scenarios involving a source horizontally displaced from a building (#1 model) or exactly beneath a building (#2 model). Those models were designed to simulate transient soil gas transport. The former was later modified by Symms et al. [42], which further included the influences of a concrete foundation by employing a crack factor.

In several models developed from CSOIL [33], like VOLASOIL [44, 45] and CSOIL 2000 [46], the indoor depressurization was considered to create an upward advection from the source to the building. Like the OCHCA VI model, these models also assume a negligible contaminant concentration in the crawl and indoor space to obtain the contaminant upward flow rate in a steady-state. Those models are all based on Jury's pesticide model [4–7] except for no biodegradation. Murphy and Chan [43] also proposed a model with coupled transport involving diffusion and uniform advection. Their work can predict mass exchange between multimedia environmental compartments with mass transfer coefficients. In specific applications, the whole model can be simplified as a three-compartment structure, similar to that of the J-E model [18].

In lateral soil gas transport, advection can be the primary driving force in the presence of significant pressure differences. For example, in the laterally located landfill case (#3 model of Little et al. [30]), a considerable pressure difference between a landfill site and the surface air was considered to simulate lateral transport of landfill gas from the landfill site to the target building. In most VI scenarios of interest, such significant pressure differences are not expected. Most VI models focus mainly on vertical transportation without significant horizontal pressure gradients. Only in a few cases may a vertical soil gas advection be caused by the bubbles generated in the source zone. To simulate the mass transfer of contaminants from the NAPL source to entrapped air in the vadose zone, a mathematical solution was developed by Ma et al. [71] based on bubble-facilitated VOC transport. They reported that the mass flux of bubble-facilitated VOC transport could be of two orders of magnitude higher than that of diffusion-limited VOC transport.

1.2.1.2 Two-Dimensional (2D) Screening Models Involving No Biodegradation

Theoretically, it is possible for a model with two or three dimensions to simulate heterogeneous conditions at the upper boundary by considering soil gas transport in both horizontal and vertical directions. However, the difficulty increases significantly for obtaining an analytical solution in a multidimensional scenario. Therefore, to reduce the difficulties in developing a 2D analytical model, it is necessary to use simplified boundary conditions based on a basic transport mechanism. A common approach is to assume a diffusion-dominated soil gas transport that can be simulated with the Laplace equation. This assumption is acceptable for VI modeling as we discuss above that advection only plays a role in a small high-permeability zone near the building foundation [2].

For example, Yao et al. [22] used the Schwarz-Christoffel conformal transform [72] to develop an analytical equation to approximate the perimeter subslab concentration. Their solution was obtained by assuming the vapor source is uniform, of infinite size and beneath a building with an impenetrable foundation slab. As mentioned above, the Laplace equation was employed as the governing equation by only considering soil gas diffusion in uniform soil. In Yao et al.'s work, the soil gas transport was treated as independent of the entry into the structure. Three-dimensional numerical studies also showed this assumption that the advection through cracks in the foundation slab would not affect the general soil gas concentration profiles [21, 73]. With the approximated subslab crack concentration, the indoor air concentration can be calculated with the same approach employed in the J-E model.

Based on the work of Yao et al. [22], Shen et al. [36] extended the mathematical solution to simulate the 2D subfoundation soil gas concentration profile for cases with uniform soil properties and moisture content. In this way, it provides an opportunity for the first time to use measured soil gas concentration to verify the simulated results. In 2016, Shen et al. [74] modified the mathematical solution to include a finite but uniform vapor source. Later, Yao et al. [40] in 2017 further improved the model developed by Shen et al. to develop CVI2D, a 2D chlorinated VI model involving vertical soil heterogeneity. A significant improvement of this model is that it can work practically for a common groundwater source, including vertical moisture heterogeneity caused by the capillary fringe effect. The simulated results by CVI2D were compared with measured soil gas profiles in a case involving layered soil, and they were reported in general agreement. Also, for a basic VI scenario involving a uniform distributed vapor source and a building surrounded by open groundwater surface, the CVI2D can do a job as good as numerical models by simulating steady-state soil gas concentration profiles and predicting indoor air concentrations. Moreover, by including the van Genuchten equation to simulate the water retention curve, the risks predicted by CVI2D were less conservative by 1–2 orders of magnitude than the results by the EPA implementation of the J-E model, which is because the latter adopted a two-layer approach (capillary fringe and vadose zone) [68]. In a soil layer, the moisture content is assumed as uniform.

To investigate the influences of lateral source-building separation distances, Lowell and Eklund [38] were the first to introduce a 2D analytical model to quantify the effects of the horizontal separation distance between the source and the target building. Their model predicts an exponential decay in soil gas concentration with the lateral transport distance. When the vapor source is located close to the ground surface, the decay rate is significant due to the fast escape of contaminants through the soil surface during the lateral transport. In Lowell and Eklund's model, the foundation's influences were not included in soil gas transport for simplifying the calculation. This issue was addressed later by Yao et al. [22], which includes the influences of building foundation on subslab concentration near a perimeter crack [34]. Though the blocking effect of the impermeable foundation was included, the predictions by the modified model were similar to those by Lowell and Eklund [38], as the lateral source-building separation distance plays a dominant role in determining the soil gas concentration attenuation. In a more recent study, this method was further modified to include the role of nonuniform geology, such as layering and surface pavements [35], and later examined by a pilot-scale tank experiment [75, 76]. Based on the same scenario, a 2D semi-analytical solution for the lateral transport was also proposed by Feng et al. [41], which reported that the width of the lateral inclusion zone increases logarithmically and linearly with source concentration and depth, respectively.

In the above 2D models, a vapor concentration boundary condition was applied at the upper edge of the source. However, McHugh et al. [39] in 2003 claimed such an approach might cause an over-conservative estimation of contaminant flow rate into the building for a groundwater source. They suggested that the soil gas release rate from the groundwater source determines the upward diffusion rate and then the entry rate. They introduced a screening-level model based on the groundwater release rate to calculate the screening level of groundwater concentration in steady-state scenarios [2].

1.2.2 The Role of Biodegradation in VI

At VI sites contaminated by petroleum products, biodegradation can enhance the soil gas concentration attenuation with the presence of suitable microbes and oxygen [33, 34, 37–39, 42–44], and some screening models have included biodegradation effects in predicting contaminant indoor air concentration under such circumstances

[63, 77]. In fact, US EPA recommended including the oxygen-limited biodegradation for evaluating VI of petroleum hydrocarbons [78].

First, it is important to consider what the term "biodegradation" of a contaminant really entails. Does it involve the compound's complete mineralization or conversion to other more or less hazardous species? In the latter case, for example, trichloroethylene (TCE) can undergo complex multistep biodegradation leading to other species of concern such as vinyl chloride. Another, for instance, biodegradation of TCE is inherently anaerobic [45]. In addition, biodegradation rates can vary widely; those for chlorinated solvents such as TCE are generally much slower than those for petroleum compounds in the absence of other agents that can enhance biodegradation. For example, TCE biodegradation can be significantly improved in the presence of methane-oxidizing bacteria and a source of methane [46].

To understand the possible role of biodegradation in any particular scenario, it is thus first necessary to know many details regarding the chemistry of the process. For example, it needs to be known whether the contaminant itself is the main substrate (food source) for the bacteria, or if there is some other substrate that supports the microbial growth such that the degradation of the contaminant of interest is merely a side-process. Whether or not the main catabolic (microbial energy-producing) processes are inhibitory to contaminant degradation might not always be clear.

The situation is also complicated by myriad other factors outside of what the contaminant and microbial populations of interest might be. For example, soil pH is often a key factor, as petroleum compound degradation is often the highest near-neutral pH. The process is generally not very effective outside the range of pH 6–8. Water is also a key factor. Some water is needed for microbe growth, but soil saturation can be harmful to aerobic degradation by slowing down oxygen diffusion to the reaction zone. Soil temperature can affect biodegradation rates through some kind of Arrhenius dependence. The presence of soil carbon can deter biodegradation by tightly adsorbing contaminant species, such that they are unavailable to microbes [79]. Thus, it is fair to conclude that any time a simple empirical rate law is offered for describing biodegradation, it must automatically be viewed with some suspicion. In modeling VI processes in the presence of biodegradation, it is often only possible to incorporate simple empirical rate laws, as the full details of the chemical processes are not known, or if some features are known, then many of the needed parameters may not be available.

1.2.3 Soil Gas Transport Involving Biodegradation

The first question that is considered here is whether the contaminant of interest is the organic matter that supports the growth of the microbes. The cases where some other organic matter, besides the contaminant of concern, supports growth has been considered by Schmidt et al. [80] For example, if the microbial population density is near a maximum value (as determined by another substrate). The contaminant concentration is low relative to that of other substrates, a simple first-order degradation rate will be observed:

$$-\frac{dS}{dt} = k_1 S \tag{1.10}$$

where S is the contaminant concentration and k_1 the effective rate constant. On the other hand, if the microbial population is near a maximum, but the contaminant concentration is also high, an apparent zero-order biodegradation rate would be observed for the contaminant:

$$-\frac{dS}{dt} = k_0 \tag{1.11}$$

Several other forms were also considered, along with models in which the timescales for microbial growth played a role in determining the timescale for contaminant biodegradation.

If the contaminant itself is the substrate that limits microbial growth, then the Monod equation needs to be considered [73, 81]. At low concentrations of a substrate (i.e., a hydrocarbon food source), the microbial population is small. With increasing substrate concentrations, the microbial population itself grows along with the population growth rate until a maximum growth rate is reached.

The expression governing the growth rate of microbes in contact with a substrate is:

$$\mu = \frac{\mu_{\max} S}{S + K_{\mathrm{m}}} \tag{1.12}$$

where μ is the specific growth rate coefficient [$M_{mic}/L^3/T$] of the microbial population, $\mu_{\max}$ is the maximum growth rate coefficient [$M_{mic}/L^3/T$], S is the concentration of limiting substrate [M_{sub}/L^3], and K_m is Monod coefficient [M_{sub}/L^3] defined as the value of S when $\mu = 0.5\mu_{\max}$.

If the growth of the microbial population is based on the consumption of the limiting substrate, i.e., the biodegradation of contaminants in the soil,

$$-\frac{dS}{dt} = \frac{V_{\max} S}{S + K_{\mathrm{m}}} \tag{1.13}$$

where $V_{\max}$ is the maximum biodegradation rate of the limiting substrate/contaminant [$M_{sub}/L^3/T$].

Based again on the Monod equation, zero-order and first-order rates may be found for limiting cases. When S is large, zero-order with respect to S is approached, and the microbial population growth does not depend on the amount of substrate. Thus, a zero-order rate contaminant biodegradation is obtained when the limiting substrate concentration is much greater than K_m.

$$\text{If } S'' K_{\mathrm{m}} \Rightarrow \mu = \mu_{\max}, \text{ then } \frac{dS}{dt} = -k_0 \tag{1.14}$$

where k_0 is the zero-order reaction rate constant [$M_{substrate}/L^3/T$].

This says that substrate reaction rate (now taken to be contaminant destruction rate) is zero order in its concentration because something other than substrate availability begins to limit the process (e.g., the microbe population reaches a steady maximum value).

When S is small, first-order dependence on S is observed, and the growth of the microbial population is limited by the availability of substrate.

First-order rate bio-reaction leads to an exponential decay of substrate concentration with time

$$\text{If } S'' K_{\mathrm{m}} \Rightarrow \mu = \frac{\mu_{\max} S}{K_{\mathrm{m}}}, \text{ then } \frac{dS}{dt} = -k_1 S \tag{1.15}$$

and thus

$$\frac{S}{S_0} = \exp\left(k_1 t\right) \tag{1.16}$$

where k_1 is the first-order reaction rate constant [1/*T*]. The microbe population will grow in proportional to substrate concentration, but it consumes the substrate, and the rate of substrate loss is itself proportional to substrate concentration, in this case.

The role of oxygen in biodegradation has not been discussed in detail, but the importance of oxidizer supply cannot be underestimated. First, it should be said that the oxidizer, or electron acceptor, in a biodegradation process can be more than just molecular oxygen. Other oxidized species such as nitrate, iron (III) oxides, and sulfates can fulfill that role, and even carbon dioxide can contribute oxygen in a process that has another significant energy source to drive it. Thus, again, the potential complexity of biodegradation processes shows itself.

Borden and Bedient [82] suggested that the rate of substrate removal should be represented by an expression that explicitly recognized the likely dominant role of oxygen:

$$\frac{dS}{dt} = -MV_{\max}\left(\frac{S}{K_{\mathrm{m}} + S}\right)\left(\frac{O}{K_{\mathrm{o}} + O}\right) \tag{1.17}$$

In which M represents the microbe population, O represents the oxygen concentration, and K_o represents the half-saturation constant for oxygen, analogous to K_{m} for the contaminant substrate. A similar expression may, of course, be written for oxygen consumption:

$$\frac{dO}{dt} = -MFV_{\max}\left(\frac{S}{K_{\mathrm{m}} + S}\right)\left(\frac{O}{K_{\mathrm{o}} + O}\right) \tag{1.18}$$

where F is the relevant stoichiometric coefficient relating oxygen to substrate consumption.

It is apparent that all of the above-discussed treatments of the limiting cases of the Monod equation can also be applied here, and so the oxygen dependence may be

Table 1.3 Summary of biodegradation kinetic expressions used in VI screening models

Models	Reaction kinetics
Sanders and Stern [52] /(#2 model)	First-order reaction
Anderson et al. [57]	First-order reaction
T & R model [58]	First-order reaction
VIM [51]	First-order reaction
Ostendorf and Kampbell [60]	Monod reaction
Johnson et al. [20]	First-order reaction
Roggemans et al. [62]	Instant reaction, coupled O_2
Parker [64]	Zero-order reaction, coupled O_2
Devaull [63]	Piecewise first-order reaction, coupled O_2
Verginelli and Baciocchi [65]	First-order reaction for both aerobic and anaerobic, coupled O_2
PVI2D [66]	Piecewise first-order reaction, coupled O_2

Adapted with permission from [2]. Copyright 2013 American Chemical Society

zero or first order in oxygen or some complex function as implied by the Monod form. In considering the sensitivity of the solution of these equations in situations in which diffusion, dispersion, and convection determine the relevant species transport rates, Borden and Bedient [82] also concluded that an instantaneous reaction assumption was often appropriate for describing the concentration distributions between a hydrocarbon substrate source and an oxygen source. This assumption has subsequently been used elsewhere to describe the concentration profiles in vapor intrusion cases (see below).

There is one final aspect to the above analysis that bears brief mention. The movement of microbes might also need to be taken into account. The growth of the microbe population itself is again described by equations having the form:

$$\frac{dM}{dt} = -MFV_{\max}\left(\frac{S}{K_{\mathrm{m}} + S}\right)\left(\frac{O}{K_{\mathrm{o}} + O}\right) + k_{\mathrm{c}}YC - bM \tag{1.19}$$

where Y represents the microbial yield coefficient (g cells/g substrate), k_c represents the decay rate of natural organic carbon (not the contaminant substrate), C is the natural organic carbon concentration in the soil, and b is the microbial decay rate constant. This expression explicitly allows for the role of other naturally occurring substrates, as discussed above, as well as a process of microbe removal by some decay process. This equation would also need to be solved subject to a transport equation that allows for diffusion, dispersion, and advection of microbial species. Again, Borden and Bedient [82] considered examples where this was done, but in general, this is beyond what would be included in normal VI modeling.

As described below, Yao et al. [2] also summarized the screening models involving biodegradations based on different forms of governing equations. Table 1.2 summarizes the governing equations of VI risk screening models involving biodegradation, the term of which is shown as "R_i". Table 1.3 shows the different

Table 1.4 Summary of assumed biodegradation rate constants in VI models

Reaction kinetics	Models	Rate constants
Monod equation $\frac{dS}{dt} = \frac{V_{\max}S}{S+K_m}$	Ostendorf and Kampbell [60]	Hydrocarbons: $V_{\max} = (5.56\sim11.3) \times 10^{-9}$kg/m^3/s $K_m = (5.56\sim8.67) \times 10^{-4}$kg/m^3
Zero-order reaction $\frac{dS}{dt} = -k_0$	VADBIO [83]	BTX and fuel hydrocarbons $k_0 = (1.56\sim3.78) \times 10^{-7}$kg/m^3/s
	Parker [64]	$k_0 = 1\sim10$ mg/kg/day (maximum decay rate)
First-order reaction $\frac{dS}{dt} = -k_1 S$	R-UNSAT [48]	k_1(MTBE) $= 1 \times 10^{-8}$s^{-1} k_1(BTEX) $= (1\sim100) \times 10^{-7}$s^{-1}
	IMPACT [84–86]	k_1(VOC) $= 4.46 \times 10^{-8}\sim5.01 \times 10^{-7}$s^{-1}
	VADBIO [83]	BTX and fuel hydrocarbons $k_1 = (1.39\sim3.33) \times 10^{-4}$s^{-1}
	T & R model [58]	k_1(Benzene) $= 1.93 \times 10^{-8}$s^{-1} in soil k_1(Benzene) $= 5.99 \times 10^{-7}$s^{-1} in air
	The ASU model [87, 88]	k_1(BTEX) $= (0.018\sim2) \times$ h^{-1} in water
	Devaull [63, 65]	Aromatic hydrocarbons $k_1 = 9.14 \times 10^{-6}$s^{-1} in water Straight chain and branched aliphatic hydrocarbons $k_1 = 8.22 \times 10^{-4}$s^{-1} in water
	Verginelli and Baciocchi [65]	BTEX and other hydrocarbons $k_1 = 0.79 - 71$ h^{-1} for aerobic biodegradation $k_1 = 1.9 \times 10^{-4} - 1.55 \times 10^{-2}$h^{-1} for anaerobic biodegradation
Second-order reaction $\frac{dS}{dt} = -k_2 S[O]$ Where $[O]$ is the oxygen concentration	The Brown model [89]	k_2(Benzene) $= (1.79\sim179) \times 10^{-5}$m^3/kg/s in water

Note: *BTEX:* benzene, toluene, ethylbenzene, and xylene; *BTX:* benzene, toluene, and xylene; *MTBE*: Methyl tert-butyl ether
Reprinted with permission from [2]. Copyright 2013 American Chemical Society

forms of R_i employed in those screening models, and Table 1.4 shows some of the reaction rate constants employed in screening and numerical models. Models with labels of "1 species" involves only kinetics of substrate (contaminant) loss, while those labeled "2 species" include coupled transport and reaction of oxygen and contaminant, assuming biodegradation only occurs in aerobic conditions.

1.2.3.1 1D Screening Models Considering Biodegradation

For models involving biodegradation, it is a challenge but not important to obtain an analytical solution, especially when only one species (i.e., the contaminant) is considered. For coupled biodegradation and transport of multiple species, it is more common to be solved with numerical codes. In 2013, Yao et al. [2] summarized

VI screening models including biodegradation based on the employed governing equations, as shown below.

The #3 model developed by Little et al. [30] was employed in Sanders and Stern [52] to include an empirical reaction rate, first order to contaminant concentration, representing contaminants' degradation. In Jeng et al. [50], diffusion, but not advection as employed in Little et al., was considered as the dominant lateral transport of contaminants, and first-order biodegradation is included. A similar approach was employed by Ririe and Sweeney [47], which obtained the contaminant soil gas concentration profile by assuming aerobic biodegradation in the whole unsaturated soil. Johnson et al. [20] also employed the same reaction kinetics during the soil gas transport to include biodegradation in the J-E model. In Parker's work [64], zero-order, instead of first-order, reaction rates were employed for variable scenarios, and the contaminant flow rate into the building of concern was calculated by eliminating the degradation rate from the predicted upward diffusion rate by the J-E model. Later, Mills et al. [51] further updated the J-E model by including an additional nonzero background to develop the Vapor Intrusion Model (VIM). The same governing equation was again employed in the R-UNSAT model [48], which, however, assumed other boundary conditions at the soil surface to calculate the largest and lowest contaminant flow rate through the upper boundary.

In pesticide leaching models, biodegradation is as important as rainfall-induced advection in determining soil gas concentration profiles. Jury's pesticide model [4–7] was modified by Anderssen et al. [57] by extending the original simplified initial and upper boundary conditions. In Sanders and Stern [52], the upper boundary conditions in Jury's model [4–7] were replaced with a zone of influence to develop the #2 model, where the source was considered to be uniformly distributed in a depth range instead of at a certain depth. Turczynowicz and Robinson [58] also adapt the Jury's model to develop the T & R model, which is different from the #2 model in Sanders and Stern [52] by assuming contaminant source directly beneath the crawl space. In the T&R model, the concentration in the crawl space needs to be calculated to predict indoor air concentrations. The Jury's models were also reintroduced by Lin and Hildemann [59] to estimate the emissions from a landfill site, and in their study, the initial condition of the contaminant vapor source was calculated with a Dirac delta function.

All of the models mentioned above in this section employed the zero-order or first-order biodegradation throughout the whole domain to simplify the soil gas transport of only one species (i.e., the contaminant). However, for some contaminants such as petroleum hydrocarbons, oxygen is required for biodegradation [90, 91], and it is necessary to predict VI risks based on coupled transport and reaction of contaminant and oxygen. Oxygen is commonly assumed to be adequate in the upper soil layer and consumed by aerobic reaction on the way moving towards the vapor source at the bottom. Usually, the transport of two species in simulations is considered independent of each other, but the occurrence of the reaction depends on both.

Ostendorf and Kampbell [60] introduced oxygen transport in a VI model simulating 1D steady-state diffusion involving biodegradation. However, in their model, the oxygen part plays no role in determining the contaminant part, while in reality, the oxygen's existence is also determined by the reaction with the contaminant. Thus, it is necessary to couple oxygen transport with the contaminant in oxygen-limited cases.

Roggemans et al. first introduced the concept of the aerobic/anaerobic interface in VI [62]. They claimed that the interface depth could be estimated with a mass balance by assuming most of the chemicals and oxygen are consumed at the interface with a stoichiometric mass balance between the diffusion flux. Such simplification is valid only when the reaction rate is fast with respect to diffusion, as discussed earlier in Borden and Bedient [82].

The aerobic/anaerobic interface was later employed by Davis et al. [49] to simulate the oxygen concentration profile, which was used to compare with measured oxygen data to examine the reaction kinetics of the contaminant. In 2007, Devaull [63] further introduced the aerobic/anaerobic interface in modifying the J-E model to include oxygen-limited degradation in the aerobic soil layer (i.e., the soil layer above the interface). A steady-state analytical solution for the soil gas concentration profiles of oxygen and contaminant was obtained based on the mass continuity at the interface. It should be noted in the original version [63], the entry process through the foundation slab was not included, and contaminant indoor air concentration was considered the same as the subslab concentration, which was calculated based on the mass balance between the upward diffusion rate at the foundation slab and indoor-outdoor exchange rate. This analytical solution was later implemented in a spreadsheet named BioVapor [92], where the solution was further modified to include the influences of foundation, as the J-E model assumed. In a model developed by Verginelli and Baciocchi [65] both anaerobic and aerobic degradation were included. In their model, there is the generation of methane in the anaerobic reaction. Later in 2014, they modified the model developed by DeVaull [63] by considering the influence of building footprint size in determining the depth of the aerobic/anaerobic interface [93]. As a result, their predictions are in good agreement with Abreu and Johnson [87].

The fermentation of ethanol-blended fuel release can generate methane and build up soil gas pressure in the source zone. As a result, an upward advective soil gas flow could be caused, accompanied by a higher health or even explosion risk. To address the problem, Yao et al., in 2015, developed a PVI model (i.e., MVI) involving both an upward advection and a piecewise first-order oxygen-limited degradation [67]. The authors reported that high vapor concentration of the methane source could cause an oxygen shadow in the subfoundation with the help of upward advection, which only plays a role in increasing volumetric soil gas entry rate into the building under such circumstances.

1.2.3.2 2D Model Involving Biodegradation

The only available 2D screening/analytical model involving oxygen-limited biodegradation is PVI2D, developed by Yao et al. in 2016 [66]. Based on the Laplace equation and the Schwarz-Christoffel mapping method [36], PVI2D was developed based on a conceptual site scenario involving a steady-state diffusion-dominated vapor transport in a homogeneous soil and an impervious foundation slab. PVI2D could replicate the 2D soil gas concentration profiles in some primary PVI cases, as Abreu and Johnson showed in 2006 [87]. It was also reported that predicted results by PVI2D were in agreement with measured hydrocarbon and oxygen profiles by Patterson and Davis in 2009 [94]. As shown in Fig. 1.2, PVI2D was also implemented in a spreadsheet named PVI2D toolkit [95], which is available for free download via www.pvitools.net.

1.3 The Entry Process into the Constructure

Contaminant soil gas transport is the most challenging to be simulated in VI modeling, but it is not all of it. Further efforts are still needed to investigate the migration route from soil and contaminant mass transfer into the indoor air [2]. The former is necessary to better define upper boundary conditions in soil gas transport, while the latter is critical to establish the contaminant indoor air concentrations. Due to the difficulty in identifying the air change rate of the building of concern, the contaminant flow rate into the structure was proposed as an alternative to the indoor concentration [26, 27]. In recent practice, this contaminant mass flow rate, or emission rate, is employed to describe VI potential in applying the building pressure cycling technique, or control pressure method, as such application would affect the indoor air exchange rate [96].

1.3.1 The Governing Equations for the Contaminant Escape from the Soil

The escape process of contaminant from the soil is critical to illustrate the boundary condition at the soil surface. Yao et al. proposed a summary of upper boundary conditions for soil gas transport, as described below [2].

The general governing equation at the soil surface could be expressed as:

$$Ac_{\mathrm{ig}} + B\nabla c_{\mathrm{ig}}\big|_{z=0} = f(x, y, 0) \tag{1.25}$$

where A and B are constants to define the influences of advection and diffusion, respectively, (x, y, z) are positional coordinates, where z is the vertical direction, and $z = 0$ is the ground surface. f is some boundary value function of (x, y, z) [2].

The simplest boundary condition is to assume the ground surface is entirely open to indoor air where the contaminant concentration is treated as zero as that in the atmosphere. In such a case, Eq. (1.25) becomes [2]

$$c_{\mathrm{ig}}\big|_{z=0} = 0 \tag{1.26}$$

The contaminant entry into the building or any other constructure is assumed to equal that escaped from the soil:

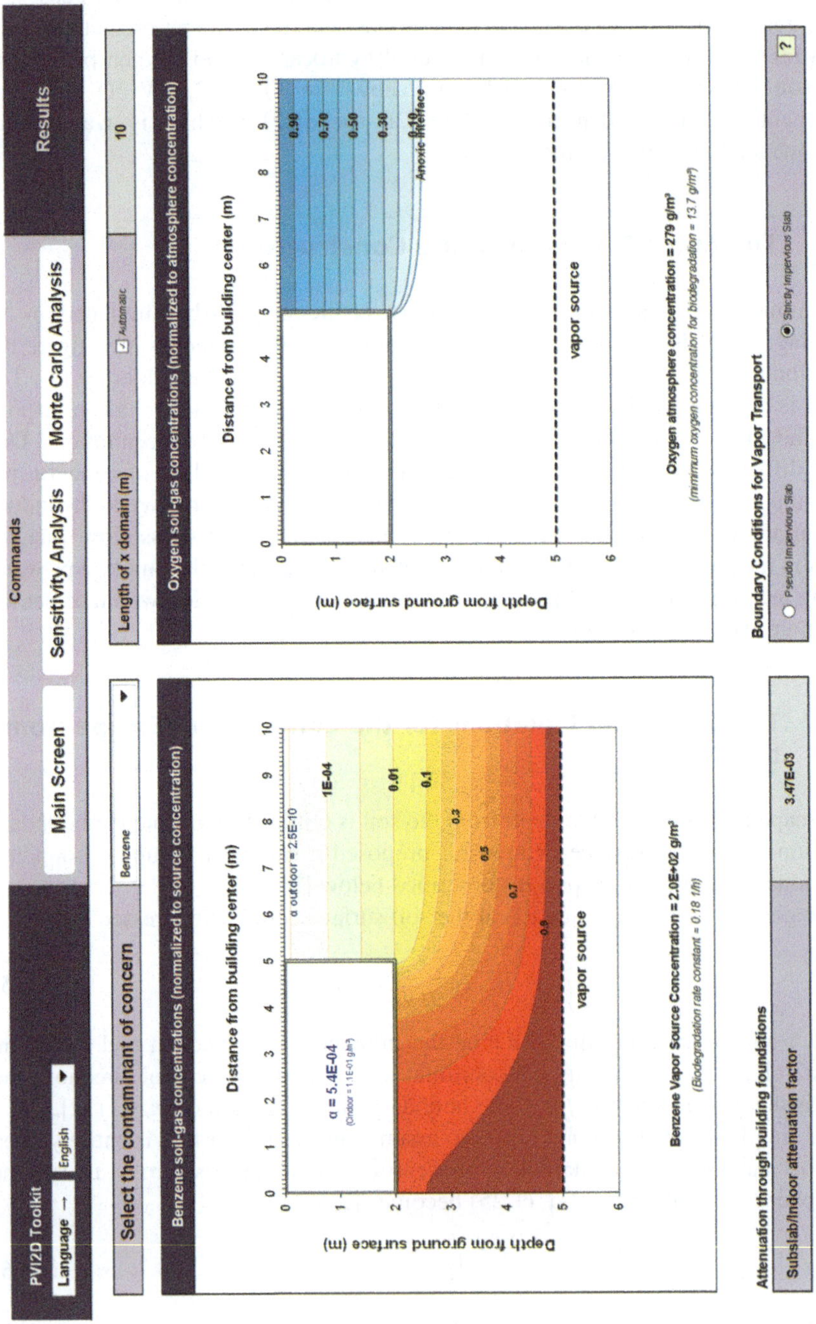

Fig. 1.2 Simulated 2D soil gas concentration profiles of hydrocarbon and oxygen with PVI2D toolkit

$$J_{\mathrm{us}} = J_{\mathrm{soil}} \tag{1.27}$$

where J_{us} and J_{soil} are the contaminant entry rate into the upper space (i.e., the enclosed space of the concerned building) and escape rate from the soil, respectively (M/T) [2].

To estimate contaminant flow rate into the constructure, the diffusive layer theory from Jury's pesticide model [4–7] was reintroduced in some VI screening models. According to the theory, a hypothetical thin layer is assumed between the soil surface and the upper space. So a diffusive flux equation can be applied as the upper boundary condition for contaminant soil gas transport. In such a way, the contaminant escape rate from the soil equals the rate across the thin layer/into the enclosed space [2].

In this case, Eq. (1.25) becomes the following form [2]

$$\nabla c_{\mathrm{ig}}\big|_{z=0} = f(x, y, 0) \tag{1.28}$$

However, this boundary condition is valid only for houses with dirty floors (i.e., no foundation slab and the soil was in direct contact with the air) but is unrealistic for typical buildings with foundation slabs that can block soil gas flow. Therefore, this equation was more commonly used for buildings with a crawl space, where the soil surface is also open to the air [2].

Nonetheless, not all of the buildings have a foundation like crawl space in the absence of a concrete or wooden slab, and it is typical for buildings with a basement or slab-on-grade construction, the foundation should not be ignored in VI modeling due to its blocking effect to soil gas transport. In the OCHCA VI model [31], an attenuation factor to the flux was introduced to represent the blocking effect due to the foundation slab, but the soil gas concentration at the soil/foundation interface was still assumed zero [2].

In the OCHCA VI model, the contaminant mass entry rate into the upper space can be expressed as:

$$J_{\mathrm{us}} = bJ_{\mathrm{soil}} \tag{1.29}$$

where b is an empirical attenuation factor to upward soil gas flux due to the presence of the slab. It should be noted that contaminant concentrations at the subslab (i.e., interface between soil and slab) are not negligible in actual VI cases shown by both field measurements and theoretical simulations [2].

An alternative to the diffusive layer theory is the permeable slab assumption, but in reality, it also depends on empirical parameters for the heterogeneous concrete slab. In Ferguson et al. [97], for example, the slab was characterized by a combination of various constructions materials located in series [2].

The most popular and maybe also the most realistic way to simulate the soil gas entry into a structure is to introduce a crack hypothesis [2]. In many studies of radon intrusion, foundation slab cracks were identified as the major entry pathway [10–13, 98]. The development of Nazaroff's equation [25] to calculate volumetric soil gas flow rate into a buried pipe was a great help in considering soil gas advection. This advection is not critical in the whole contaminant transport but is potentially quite critical around the crack, where the soil permeability is usually relatively higher than the rest of the soil domain [2]. The J-E model [18] and other models [72, 94] developed from it benefited greatly from this idea. This crack-entry assumption was also included in a series of models by Yao et al. [22, 34, 35, 40, 66].

In the presence of advection and diffusion, the general governing Eq. (1.25) can be transformed into the following form [2]

$$Ac_{\text{ig}} + \nabla c_{\text{ig}}\big|_{z=0} = \begin{cases} 0 & \text{beneath the slab area} \\ f(x,y,0) & \text{at the crack area} \end{cases} \tag{1.30}$$

Obviously, this boundary hypothesis is more appropriate when applied to multi-dimensional models, as the boundary condition assumed here allows considering at least two dimensions [2].

Nazaroff's equation [25] was modified in some versions of the J-E screening tool is:

$$Q_{\text{ck\_JE}} = \frac{2\pi k \Delta p L_{\text{ck}}}{\mu_{\text{g}} \ln\left(2d_{\text{f}}/w_{\text{ck}}\right)} \tag{1.31}$$

where $Q_{\text{ck\_JE}}$ is the volumetric soil gas flow rate into the constructure [L^3/T], k is soil permeability [L^2], Δp is the indoor-outdoor pressure difference [$M/L/T^2$], L_{ck} is the length of the perimeter crack, μ_{g} is the soil gas viscosity [M/L/T], d_{f} is the foundation depth [L], and w_{ck} is the crack width [L] [2].

Johnson [24] once recommended the ratio of volumetric rate into the building to building air exchange rate as an alternative to Nazaroff's equation [25] as employed in the J-E model [18]. The problem however is that such a value of this ratio would also be empirical [2]. A similar recommendation was also implemented in the latest EPA VI spreadsheet [68], and so is the generic subslab-to-indoor air concentration attenuation factor (0.03) recommended by EPA was obtained by the statistical analysis of the EPA's VI database [99]. However, such a value is considered over-conservative by recent studies [100].

Table 1.5 summarizes the different entry routes assumed in various analytical and numerical VI models.

1.3.2 Indoor Air Concentration Calculation

Some studies [26, 27] have suggested that due to the uncertainties in building operational conditions, contaminant indoor air concentration (or the related empirical attenuation factor) may not be suitable to indicate the VI threat and could be replaced with contaminant flow rate into the building or building loading rate. Many other inherent factors characterizing the enclosed space, other than the indoor air exchange rate, also play significant roles, such as the structure of the building.

Table 1.5 The summary of entry processes used in vapor intrusion modeling

Entry scenario	Models
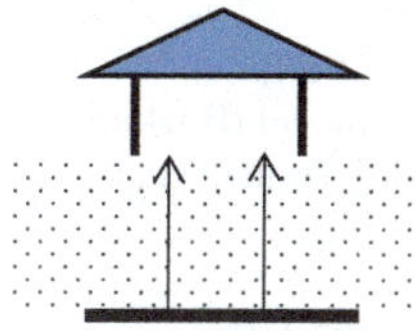 No-barrier entry	Little et al. (#1,3) [30], Devaull [63], Jeng et al. [50], Sanders and Stern (#1) [52]
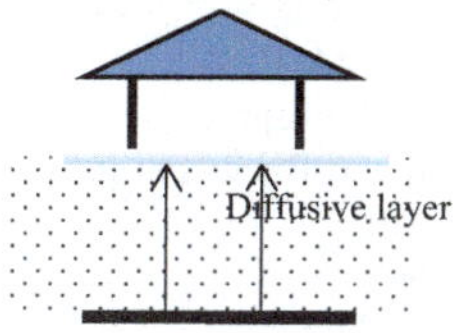 Empirical diffusive layer resistance	Little et al. (#2) [30], Sanders and Stern (#2) [52], VOLASOIL [44]
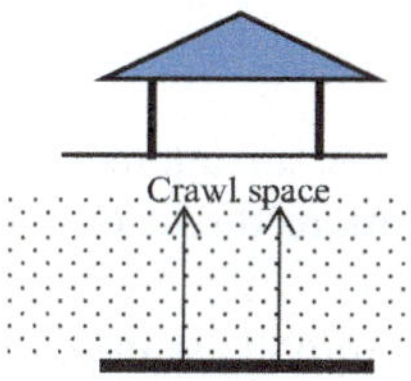 No-barrier entry into crawl space	CSOIL [33], CSOIL 2000 [46], VOLASOIL [44, 45]
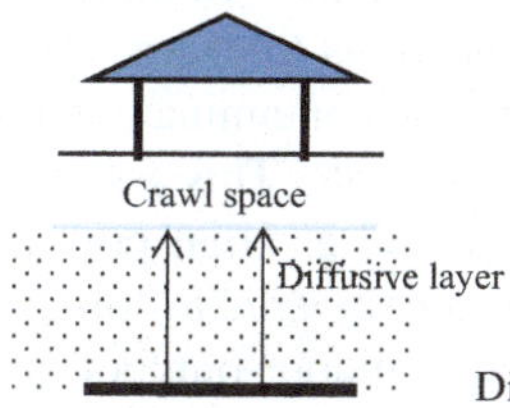 Diffusive barrier into crawl space	VOLASOIL [44], T & R model [58], IMPACT [84–86], VIM [51]
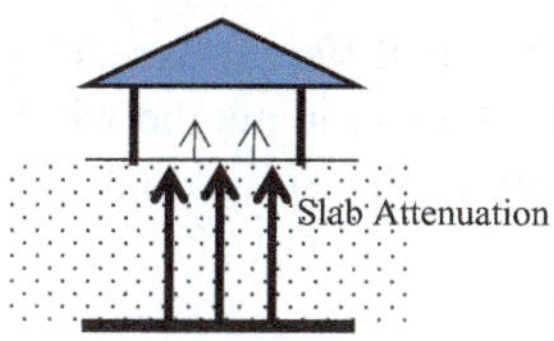 Empirical slab attenuation	OCHCA [31]
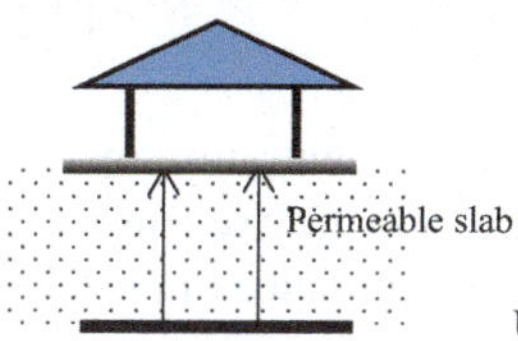 Uniformly permeable slab	Krylov and Ferguson [101], Ferguson et al. [97]
	J-E model [18], Johnson et al. [20], Park [32], Murphy and Chan [43], The ASU model (Abreu and Johnson [21, 87],

(continued)

Table 1.5 (continued)

Entry scenario	Models
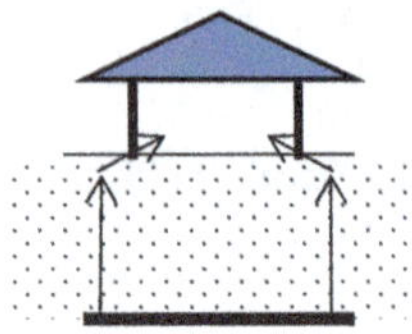 Foundation slab crack	Abreu [102], Abreu et al. [88]), The Brown model (Bozkurt et al. [79, 80], Pennell et al. [81], Yao et al. [73]), Parker [64], CompFlow Bio [103], VIM [51], VOLASOIL [44], Symms et al. [42], Verginelli and Baciocchi [65], CVI2D [40], PVI2D [66] Ma et al. [71]

Adapted with permission from [2]. Copyright 2013 American Chemical Society

However, calculations of a contaminant indoor concentration (or an attenuation factor with the respect to the subsurface source) give people a more direct understanding of the importance of the VI effect.

In the calculation of contaminant indoor air concentration, a common assumption employed by all VI screening models is to treat the indoor space as a Continuous Stirred Tank (CST) or several CSTs in parallel or series collections to characterize the building space [2]. As assumed in most screening models, the contaminant mass entry rate equals the amount out of the indoor space in a steady-state. This assumption could also be virtually valid even if involving a transient soil gas transport, as the time for indoor air mixing and exchange was relatively fast compared to soil gas transport if given constant indoor pressure conditions and indoor-outdoor air exchange rate.

Yao et al. [2] summarized the arrangements of CSTs in different screening models. The most general assumption employs just one CST to simulate the whole indoor space. Moreover, the contaminant escape rate from the soil equals the contaminant entry rate into the dwelling space.

$$c_{\text{in}} = \frac{J_{\text{us}} + V_{\text{b}}A_{\text{e}}c_{\text{atm}}}{V_{\text{b}}A_{\text{e}}} \tag{1.32}$$

where c_{in} and c_{atm} is the contaminant indoor and outdoor air concentration respectively (M/L), V_{b} is the enclosed space volume (L^3), and A_{e} is the air exchange rate ($1/T$). Here, the small soil gas entry rate contribution to the denominator of (1.28) is negligible compared to the indoor-outdoor air exchange rate. In most cases, c_{atm} is taken to be zero.

A second method is to employ two CSTs in series, with one as the dwelling space, and the other for a building foundation like the crawl space or basement between the soil and the dwelling space. As discussed above, this considers the actual characteristics of some buildings. Therefore, the air exchange rate of the crawl space and contaminant flow rate into the dwelling space are required to estimate the contaminant indoor air concentration (or any VI attenuation factor).

$$c_1 = \frac{J_{\mathrm{us}} + V_{\mathrm{b1}}A_{\mathrm{e1}}c_{\mathrm{atm}} - J_{\mathrm{us12}}}{V_{\mathrm{b1}}A_{\mathrm{e1}}} \tag{1.33}$$

$$c_2 = \frac{J_{\mathrm{us1}} + V_{\mathrm{b2}}A_{\mathrm{e2}}c_{\mathrm{atm}}}{V_{\mathrm{b2}}A_{\mathrm{e2}}} \tag{1.34}$$

where c_1 and c_2 is contaminant concentration (M/L^3) in the first and second CST, V_{b1} and V_{b2} is the enclosed space volume (L^3) of the first and second CST, A_{e1} and A_{e2} is the air exchange rate (1/T) of first and second CST, respectively, and J_{us12} is the contaminant mass flow rate (M/T) from the first CST to the second CST. This may provide more accurate predictions of contaminant indoor air concentration in buildings with separated spaces. If the enclosed space includes basement and dwelling space, the airflow between them might not be so free. However, this approach would undoubtedly introduce more trouble in identifying the additional parameters needed for calculation, such as the individual air exchange rate for each CST and the volumetric flow rate between CSTs.

Due to the different use of space inside the whole enclosed space, it may sometimes be necessary to divide the enclosed space into more than two CSTs. This is a simple extension of the previous case.

In VIM developed by Mills et al. [51], two parallel CSTs, representing a mixed building foundation, a basement and a crawl space, were assumed to connect to the third CST, the dwelling space.

$$c_{in} = \frac{J_{\mathrm{us13}} + J_{\mathrm{us23}} + V_{\mathrm{b}}A_{\mathrm{e}}c_{\mathrm{atm}}}{V_{\mathrm{b}}A_{\mathrm{e}}} \tag{1.35}$$

where J_{us13} and J_{us23} are the contaminant mass flow rates [M/T] from the first and second CST, respectively, into the third CST [2].

In general, the indoor air concentration is calculated based on the connectivity of CSTs, and the critical problem is obtaining enough parameters to characterize the airflow situation [2].

Table 1.6 summarizes some of the various assumptions made in screening models.

1.4 The Limits of VI Screening Tools

In practice, one of the major foci is the contaminant indoor air concentration (and related concentration attenuation factors) [26, 99], which is the direct indicator of VI threat to human health. The common structure of typical 1D screening models is based on the mass balance between the upward diffusion rate and the entry rate in the building of concern, if without biodegradation [18, 31–33, 43–45]. In these models, building operational conditions are included to simulate the soil gas transport in the

Table 1.6 The summary of indoor air concentration calculation assumptions in screening and numerical models

Entry scenario	Models
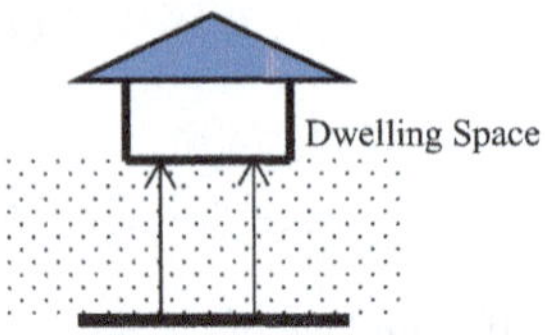 Single CST	Little et al. [30], OCHCA [31], Sanders and Stern [52], J-E model [18], Johnson et al. [20], Jeng et al. [50], Parker [64], Devaull [63], CompFlow Bio [103], Verginelli and Baciocchi [65], The ASU model (Abreu and Johnson [21, 87], Abreu [102], Abreu et al. [88]), The Brown model (Bozkurt et al. [79, 80], Pennell et al. [81], Yao et al. [73]), CVI2D [40], PVI2D [66], Verginelli et al. [93], Ma et al. [71]
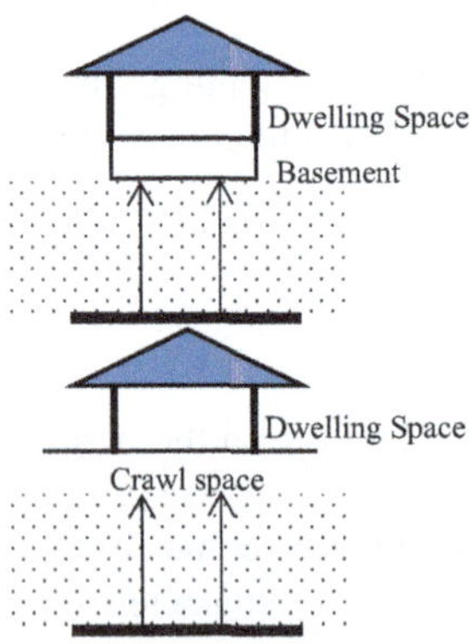 Series CSTs	T & R model [58], CSOIL [33], VOLASOIL [44, 45], CSOIL 2000 [46], IMPACT [84–86], Murphy and Chan [43], Olson and Corsi [104, 105], Ferguson et al. [97], Krylov and Ferguson [101]
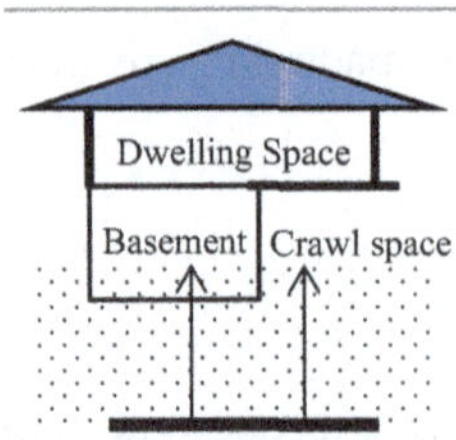 Combined series and parallel CSTs	VIM [51]

Adapted with permission from [2]. Copyright 2013 American Chemical Society

unsaturated zone, which, however, is not in agreement with common sense and the results of other sophisticated studies [21, 22]. In practice, it is believed that the soil gas advection caused by indoor-outdoor pressure difference can only affect soil gas

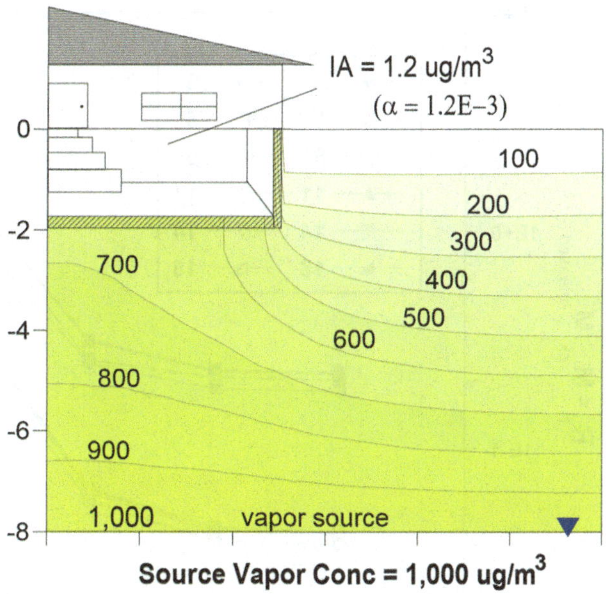

Fig. 1.3 Simulated soil gas concentration profile by the ASU model [107]

concentration profile in a small zone around the foundation crack, where the soil permeability is usually relatively higher than the rest of the soil domain.

In a classic VI conceptual scenario, proposed by Johnson and Ettinger in 1991 and employed by many other 1D models [18, 92], the contaminant soil gas enters the indoor space through the foundation crack mainly by advection [106]. The soil gas advection is usually believed to be caused by indoor-outdoor pressure differences, implying that the building of concern is surrounded by a ground surface open to the atmosphere [18, 32, 51, 106]. In such a scenario, most 1D screening models assume that the tiny foundation crack, instead of the atmosphere at the soil surface, is the final receptor of the contaminant soil gas released from the subsurface source. In some instances, such an assumption is valid as the advection caused by the indoor-outdoor pressure difference is strong enough to sweep all the contaminants beneath the building foundation into the building, but in other cases, it may not.

In most cases, the contaminant soil gas concentration profiles are generally more dominated by the diffusion from the source to the surface, with the consideration of the blocking effect by the building foundation and other potential presence of impermeable surface cover, as shown in Fig. 1.3. The atmosphere at the open ground surface, with a much larger size than the crack, plays a more dominant role of the sink to the surface soil gas. Moreover, due to the blocking effect of the impermeable building foundation, the subslab contaminant soil gas concentration is higher than that at the same depth but beneath the open soil surface, causing outward diffusion from the subfoundation to the region beyond the building footprint. Nevertheless, as we know, both vertical and horizontal soil gas transport can not be included in 1D models.

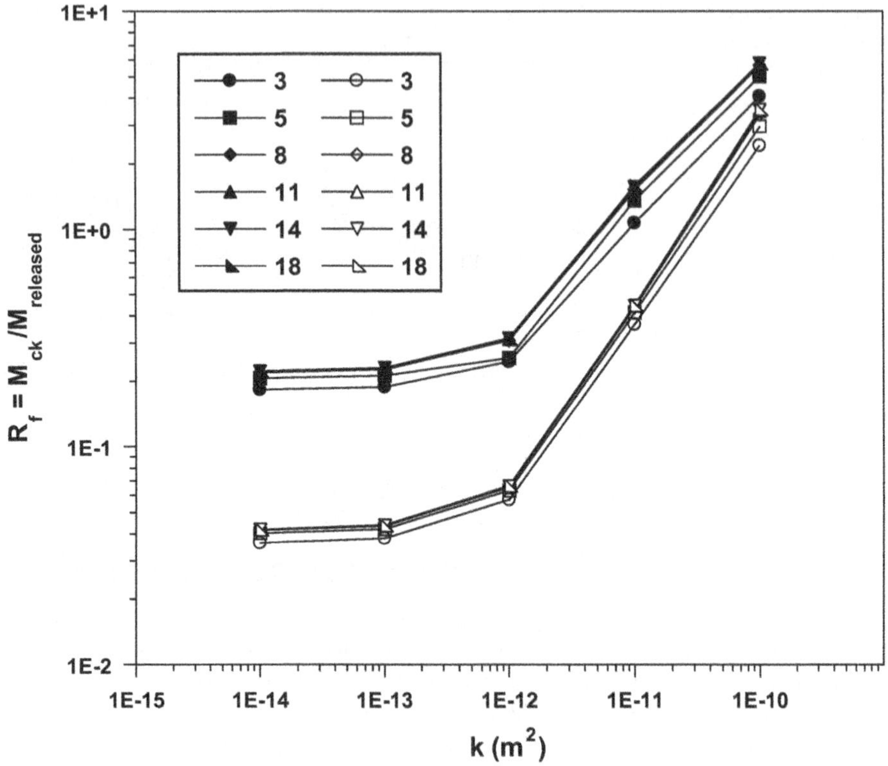

Fig. 1.4 The predicted ratio of contaminant entry rate to source release rate by the Brown model. Reprinted with permission from [73]. Copyright 2011 American Chemical Society

The forced mass balance between the upward diffusion and the entry rate into the building would likely cause a conservative prediction unless the advection is so strong that it draws contaminant soil gas beyond the building perimeter into the indoor space. The latter situation occurs only with a very porous subfoundation zone, and the source plume size is relatively large to extend beyond the building footprint. Figure 1.4 shows the predicted ratio of contaminant entry rate through the crack to that released from the vapor source of the same size as the building foundation footprint. The results suggest that in most cases with soil permeability lower than 10^{-11} m^2, the predicted contaminant entry rate is less than the source release rate, indicating a conservative prediction of 1D models involving the forced mass balance. When the soil permeability is 10^{-11} m^2, the simulated ratio is larger than 1, indicating an underestimate of the model predictions.

The available field studies are actually divided on the accuracy of the current 1D screening models. The indoor air and soil gas concentrations by the J-E model and VOLASOIL were considered to be underestimated by Hulot et al. [108], while Hers et al. [109] claimed that the calculations by the J-E model could be over-conservative by up to two orders of magnitude. In another study by Provoost et al. [110], the calculated indoor air concentrations by several 1D screening models were scattered as under or over predictions within three orders of magnitude around the measured data.

Besides the forced mass balance inherited in the 1D model, the inappropriate characterization of the vapor source could also be a possible explanation for the performance of screening models. In most modeling settings, the vapor source is always considered uniformly distributed, at least in a defined region. However, in reality, the source concentration, no matter if it is in dissolved phase or nonaqueous phase liquid, can be highly heterogeneous, both vertically and horizontally. As investigators usually rely on single measurement or measurements from single monitoring well to characterize the vapor source, it is hopeless to wish for an accurate prediction based on any modeling.

For example, to avoid the limitation of 1D screening tools, Yao et al. developed 2D analytical VI models for CVI [40] and PVI [66], namely CVI2D and PVI2D. By including a 2D conceptual scenario, it is possible to simulate both the building foundation and the open ground surface surrounding the concerned building simultaneously. Their predicted soil gas concentration profiles are in good agreement with those simulated by 3D numerical models, such as the ASU model [111]. However, like other 1D VI models, CVI2D and PVI2D still rely on the assumption involving a homogenous vapor source. As a result, its predictions of soil gas profiles are still quite different from the field observations, as shown in Fig. 1.5.

The uncertainty exists not only in the source characterization but also in building operational conditions. As discussed before, most screening models rely on an assumption of CST, without temporal and spatial variations in indoor concentrations. However, in reality, such an assumption may not work. In a long-term monitoring experiment by Holton et al. [112] and Guo et al. [96], up to three to four orders of magnitude differences were observed in indoor air concentrations measurements before indoor pressure control, and there was still one order of magnitude variation after a preferential pathway was closed. It should be noted that that building was purchased only for experimental purposes and not occupied by any residents, the presence of which would undoubtedly increase the uncertainty in building operational conditions. All of these were entirely outside the realm of VI modeling.

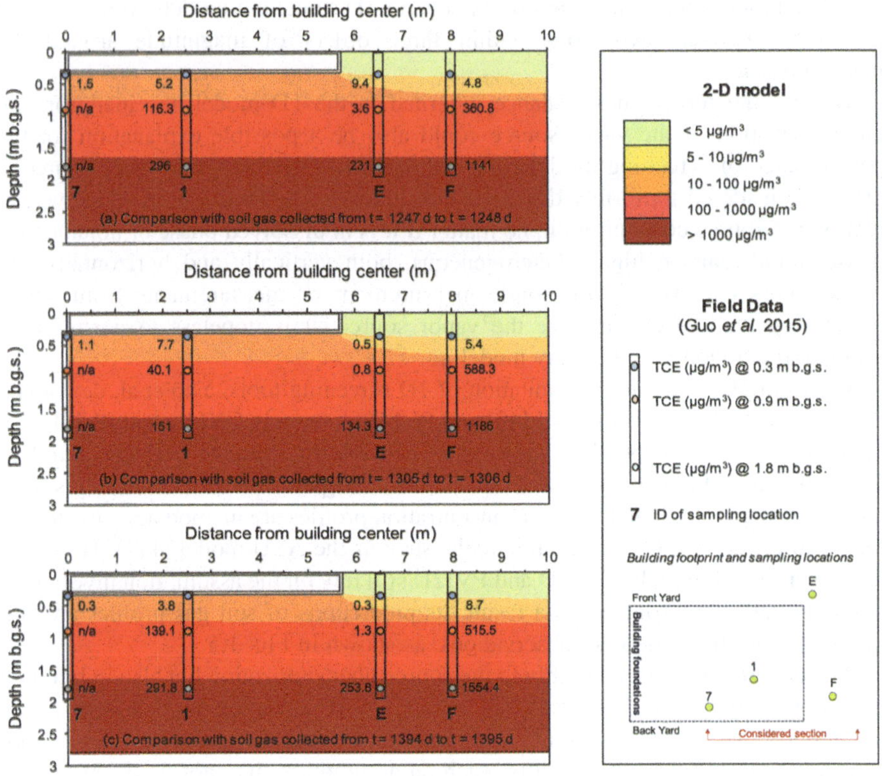

Fig. 1.5 the predicted and measured soil gas concentration profiles [40, 96]. Reprinted with permission from [40, 96]. Copyright 2017. American Geophysical Union

References

1. US EPA. Vapor intrusion. https://www.epa.gov/sites/default/files/2015-09/vaporintrusion.jpg.
2. Yao YJ, Shen R, Pennell KG, Suuberg EM. A review of vapor intrusion models. Environ Sci Technol. 2013;47(6):2457–70.
3. Verginelli I, Yao YJ. A review of recent vapor intrusion modeling work. Ground Water Monit Remediat. 2021;41(2):138–44.
4. Jury WA, Spencer WF, Farmer WJ. Behavior assessment model for trace organics in soil: I. Model description. J Environ Qual. 1983;12(4):558–64.
5. Jury WA, Farmer WJ, Spencer WF. Behavior assessment model for trace organics in soil: II. Chemical classification and parameter sensitivity. J Environ Qual. 1984;13(4):567–72.
6. Jury WA, Spencer WF, Farmer WJ. Behavior assessment model for trace organics in soil: IV. Review of experimental evidence. J Environ Qual. 1984;13(4):580–6.
7. Jury WA, Spencer WF, Farmer WJ. Behavior assessment model for trace organics in soil: III. Application of screening model. J Environ Qual. 1984;13(4):573–9.

8. Jury WA, Focht DD, Farmer WJ. Evaluation of pesticide groundwater pollution potential from standard indices of soil-chemical adsorption and biodegradation. J Environ Qual. 1987;16 (4):422–8.
9. Jury WA, Russo D, Streile G, Elabd H. Evaluation of volatilization by organic chemicals residing below the soil surface. Water Resour Res. 1990;26(1):13–20.
10. Loureiro CO. Simulation of the steady-state transport of radon from soil into houses with basements under constant negative pressure. Ph.D. Dissertation,. Berkeley, CA: University of Michigan; 1987.
11. Loureiro CO, Abriola LM, Martin JE, Sextro RG. Three-dimensional simulation of radon transport into houses with basements under constant negative pressure. Environ Sci Technol. 1990;24(9):1338–48.
12. Mowris RJ, Fisk WJ. Modeling the effects of exhaust ventilation on 222Rn entry rates and indoor 222Rn concentrations. Health Phys. 1988;54(5):491–501.
13. Nazaroff WW, Feustel H, Nero AV, Revzan KL, Grimsrud DT, Essling MA, Toohey RE. Radon transport into a detached one-story house with a basement. Atmos Environ. 1985;19(1):31–46.
14. Lindstrom FT, Boersma L, Gardiner H. 2,4-D diffusion in saturated soils. A mathematical theory. Soil Sci. 1968;106(2):107–13.
15. Oddson JK, Letey J, Weeks LV. Predicted distribution of organic chemicals in solution and adsorbed as a function of position and time for various chemicals and soil properties. Soil Sci Soc Am J. 1970;34(3):412–7.
16. Clements WE, Wilkening MH. Atmospheric pressure effects on 222Rn transport across the Earth-air interface. J Geophys Res. 1974;79(33):5025–9.
17. Schery SD, Gaeddert DH, Wilkening MH. Factors affecting exhalation of radon from a gravelly sandy loam. J Geophys Res-Atmos. 1984;89(D5):7299–309.
18. Johnson PC, Ettinger RA. Heuristic model for predicting the intrusion rate of contaminant vapors into buildings. Environ Sci Technol. 1991;25(8):1445–52.
19. Provoost J, Tillman F, Weaver J, Reijnders L, Swartjes F. Vapour intrusion into buildings – a literature review. In: Daniels JA, editor. Advances in environmental research, vol. 5. Hauppauge, NY: Nova Science Publshers, Inc.; 2010. p. 1–43.
20. Johnson PC, Kemblowski MW, Johnson RL. Assessing the significance of subsurface contaminant vapor migration to enclosed spaces: Site-specific alternatives to generic estimates. J Soil Contam. 1999;8(3):389–421.
21. Abreu LDV, Johnson PC. Effect of vapor source - building separation and building construction on soil vapor intrusion as studied with a three-dimensional numerical model. Environ Sci Technol. 2005;39(12):4550–61.
22. Yao YJ, Pennell KG, Suuberg EM. Estimation of contaminant subslab concentration in vapor intrusion. J Hazard Mater. 2012;231:10–7.
23. Yao YJ, Xiao YT, Luo J, Wang GF, Ström J, Suuberg E. High-frequency fluctuations of indoor pressure: a potential driving force for vapor intrusion in urban areas. Sci Total Environ. 2020;710:136309.
24. Johnson PC. Identification of application-specific critical inputs for the 1991 Johnson and Ettinger vapor intrusion algorithm. Ground Water Monit Remediat. 2005;25(1):63–78.
25. Nazaroff WW. Predicting the rate of 222Rn entry from soil into basement of a dwelling due to pressure-driven air flow. Radiat Prot Dosim. 1988;24(1–4):199–202.
26. US EPA. Draft guidance for evaluating the vapor intrusion to indoor air pathway from groundwater and soils (subsurface vapor intrusion guidance). Washington DC: Office of Solid Waste and Emergency Response (OSWER). EPA530-D-02-004.; 2002.
27. US EPA. Draft EPA's vapor intrusion database: preliminary evaluation of attenuation factors. Washington DC: Office of Solid Waste and Emergency Response (OSWER); 2008.
28. Millington R, Quirk JP. Permeability of porous solids. Trans Faraday Soc. 1961;57(8):1200–7.
29. Atteia O, Hohener P. Semianalytical model predicting transfer of volatile pollutants from groundwater to the soil surface. Environ Sci Technol. 2010;44(16):6228–32.

30. Little JC, Daisey JM, Nazaroff WW. Transport of subsurface contaminants into buildings. Environ Sci Technol. 1992;26(11):2058–66.
31. Daugherty SJ. Regulatory approaches to hydrocarbon contamination from underground storage tanks. In: Kostecki PT, Calabrese EJ, editors. Proceedings of hydrocarbon contaminated soils and groundwater, vol. 1. Chelsea, MI: Lewis Publishers; 1991. p. 23–63.
32. Park HS. A method for assessing soil vapor intrusion from petroleum release sites: multi-phase/multi-fraction partitioning. Global Nest J. 1999;1(3):195–204.
33. Van den Berg R. Human exposure to soil contamination: a qualitative and quantitative analysis towards proposals for human toxicological intervention values. 725201011. Bilthoven, The Netherlands: RIVM; 1994.
34. Yao YJ, Shen R, Pennell KG, Suuberg EM. Estimation of contaminant subslab concentration in vapor intrusion including lateral source-building separation. Vadose Zone J. 2013;12(3):9.
35. Yao YJ, Wu Y, Tang ML, Wang Y, Wang JJ, Suuberg EM, Jiang L, Liu J. Evaluation of site-specific lateral inclusion zone for vapor intrusion based on an analytical approach. J Hazard Mater. 2015;298:221–31.
36. Shen R, Suuberg EM. Analytical quantification of the subslab volatile organic vapor concentration from a non-uniform source. Environ Model. Softw. 2014;54:1–8.
37. Wright J, Howell M. Evaluation of vapour migration modelling in quantifying exposure. In: Proceedings Enviro, vol 04. Sydney, 2004.
38. Lowell PS, Eklund B. VOC emission fluxes as a function of lateral distance from the source. Environ Prog. 2004;23(1):52–8.
39. McHugh TE, Connor JA, Ahmad F, Newell CJ. A groundwater mass flux model for groundwater-to-indoor-air vapor intrusion. In: Proceedings of the 7th international in situ and on-site bioremediation symposium, Orlando, Fla. 2003.
40. Yao YJ, Verginelli I, Suuberg EM. A two-dimensional analytical model of vapor intrusion involving vertical heterogeneity. Water Resour Res. 2017;53(5):4499–513.
41. Feng SJ, Zhu ZW, Chen HX, Chen ZL. Two-dimensional analytical solution for VOC vapor migration through layered soil laterally away from the edge of contaminant source. J Contam Hydrol. 2020;233:103664.
42. Symms KG, Lawrence KG, Wardrop DH, Vitale RJ. Modeling VOC migration and vapor intrusion into building indoor air from subsurface soil sources. In: 5th international FZK/TNO conference on contaminated soil, Maastricht, Netherlands, Kluwer Academic Publication: Maastricht, Netherlands. 1995; pp 551–556.
43. Murphy BL, Chan WR. A multi-compartment mass transfer model applied to building vapor intrusion. Atmos Environ. 2011;45(37):6650–7.
44. Waitz MFW, Freijer JI, Kreule P, Swartjes FA. The VOLASOIL risk assessment model based on CSOIL for soils contaminated with volatile compounds. 715810014. Bilthoven, The Netherlands: RIVM; 1996.
45. Bakker J, Lijzen JPA, Van Wijnen HJ. Site-specific human risk assessment of soil contamination with volatile compounds. Report 711701049. Bilthoven, The Netherlands: RIVM; 2008.
46. Brand E, Otte PF, Lijzen JPA. CSOIL 2000 an exposure model for human risk assessment of soil contamination. A model description. Report 711701054. Bilthoven, The Netherlands: RIVM; 2007.
47. Ririe T, Sweeney R. Fate and transport of volatile hydrocarbons in the vadose zone. In: Conference of petroleum hydrocarbons and organic chemicals in ground water, Ground Water Publishing Company; 1995. pp 529–542.
48. Lahvis MA, Baehr AL. Documentation of R-UNSAT, a computer model for the simulation of reactive, multispecies transport in the unsaturated zone. Reston, VA: US Department of the Interior, US Geological Survey; 1997.
49. Davis G, Rayner J, Trefry M, Fisher S, Patterson B. Measurement and modeling of temporal variations in hydrocarbon vapor behavior in a layered soil profile. Vadose Zone J. 2005;4 (2):225–39.

50. Jeng CY, Kremesec VJ, Primack HS. Models of hydrocarbon vapor diffusion through soil and transport into buildings. In: Petroleum hydrocarbons and organic chemicals in ground water: prevention, detection and remediation conference. Houston, TX: National Ground Water Association Publishing Company; 1996; pp. 319–338.
51. Mills WB, Liu S, Rigby MC, Brenner D. Time-variable simulation of soil vapor intrusion into a building with a combined crawl space and basement. Environ Sci Technol. 2007;41 (14):4993–5001.
52. Sanders PF, Stern AH. Calculation of soil cleanup criteria for carcinogenic volatile organic compounds as controlled by the soil-to-indoor air exposure pathway. Environ Toxicol Chem. 1994;13(8):1367–73.
53. Ünlü K, Kemblowski MW, Parker JC, Stevens D, Chong PK, Kamil I. A screening model for effects of land-disposed wastes on groundwater quality. J Contam Hydrol. 1992;11 (1–2):27–49.
54. Van Genuchten MT, Wierenga PJ. Mass transfer studies in sorbing porous media I. Analytical solutions. Soil Sci Soc Am J. 1976;40(4):473–80.
55. Van Genuchten MT, Wierenga PJ. Mass transfer studies in sorbing porous media: II. Experimental evaluation with tritium ($3H_2O$). Soil Sci Soc Am J. 1977;41(2):272–8.
56. Van Cenuchten MT, Wierenga PJ, Oconnor GA. Mass transfer in sorbing porous media. III. Experimental evaluation with 2.4,5-T. Soil Sci Soc Am J. 1977;41(2):278–85.
57. Anderssen RS, De Hoog FR, Markey BR. Modelling the volatilization of organic soil contaminants: Extension of the Jury, Spencer and Farmer Behaviour Assessment Model and solution. Appl Math Lett. 1997;10(1):31–4.
58. Turczynowicz L, Robinson N. A model to derive soil criteria for benzene migrating from soil to dwelling interior in homes with crawl spaces. Hum Ecol Risk Assess. 2001;7(2):387–415.
59. Lin JS, Hildemann LM. A nonsteady-state analytical model to predict gaseous emissions of volatile organic compounds from landfills. J Hazard Mater. 1995;40(3):271–95.
60. Ostendorf DW, Kampbell DH. Biodegradation of hydrocarbon vapors in the unsaturated zone. Water Resour Res. 1991;27(4):453–62.
61. Roggemans S. Natural attenuation of hydrocarbon vapors in the vadose zone. Ph.D. Dissertation, Arizona State University, M.S. 1998.
62. Roggemans S, Bruce CL, Johnson PC, Johnson RL. Vadose zone natural attenuation of hydrocarbon vapors: an emperical assessment of soil gas vertical profile data. Washington D.C.: American Petroleum Institute; 2001.
63. DeVaull GE. Indoor vapor intrusion with oxygen-limited biodegradation for a subsurface gasoline source. Environ Sci Technol. 2007;41(9):3241–8.
64. Parker JC. Modeling volatile chemical transport, biodecay, and emission to indoor air. Ground Water Monit Remediat. 2003;23(1):107–20.
65. Verginelli I, Baciocchi R. Modeling of vapor intrusion from hydrocarbon-contaminated sources accounting for aerobic and anaerobic biodegradation. J Contam Hydrol. 2011;126 (3–4):167–80.
66. Yao YJ, Verginelli I, Suuberg EM. A two-dimensional analytical model of petroleum vapor intrusion. Water Resour Res. 2016;52(2):1528–39.
67. Yao YJ, Wu Y, Wang Y, Verginelli I, Zeng T, Suuberg EM, Jiang L, Wen YZ, Ma J. A petroleum vapor intrusion model involving upward advective soil gas flow due to methane generation. Environ Sci Technol. 2015;49(19):11577–85.
68. US EPA. EPA Spreadsheet for Modeling Subsurface Vapor Intrusion. 2017. https://www.epa.gov/vaporintrusion/epa-spreadsheet-modeling-subsurface-vapor-intrusion.
69. BP. Risc v4.03-risk assessment model for soil and groundwater applications. Sunbury: BP Oil International; 2001.
70. GroundwaterSoftware.com. RBCA Tool kit for Chemical Releases V2.6, 2.6; 2021.
71. Ma E, Zhang YK, Liang XY, Yang JH, Zhao YQ, Liu XY. An analytical model of bubble-facilitated vapor intrusion. Water Res. 2019;165:114992.
72. Carslaw HS, Jaeger JC. Conduction of heat in solids. Oxford: Clarendon Press; 1959.

73. Yao YJ, Shen R, Pennell KG, Suuberg EM. Comparison of the Johnson-Ettinger vapor intrusion screening model predictions with full three-dimensional model results. Environ Sci Technol. 2011;45(6):2227–35.
74. Shen R, Suuberg EM. Impacts of changes of indoor air pressure and air exchange rate in vapor intrusion scenarios. Build Environ. 2016;96:178–87.
75. Wang GF, Ma SS, Strom J, Suuberg E, Yao YJ, Zeng LZ. Investigating two-dimensional soil gas transport of trichloroethylene in vapor intrusion scenarios involving surface pavements using a pilot-scale tank. J Hazard Mater. 2019;371:138–45.
76. Wang GF, Xiao YT, Zuo JP, Wang Y, Man J, Tang W, Chen Q, Ma SS, Yao YJ. Physically simulating the effect of lateral vapor source-building separation on soil vapor intrusion: Influences of surface pavements and soil heterogeneity. J Contam Hydrol. 2020;235:8.
77. Johnson PC, Ettinger RA, Kurtz JP, Bryan R, Kester JE. Empirical assessment of ground water-to-indoor air attenuation factors for the CDOT-MTL Denver site. Ground Water Monit Remediat. 2009;29(1):153–9.
78. US EPA. Technical guide for addressing petroleum vapor intrusion at leaking underground storage tank sites. Washington DC: Office of Underground Storage Tanks. EPA 510-R-15-001.; 2015.
79. Bozkurt O. Investigation of vapor intrusion scenarios using a three-dimensional numerical model. Ph.D. Dissertation,. Providence, RI: Brown University; 2009.
80. Bozkurt O, Pennell KG, Suuberg EM. Simulation of the vapor intrusion process for nonhomogeneous soils using a three-dimensional numerical model. Ground Water Monit Remediat. 2009;29(1):92–104.
81. Pennell KG, Bozkurt O, Suuberg EM. Development and application of a three-dimensional finite element vapor intrusion model. J Air Waste Manage Assoc. 2009;59(4):447–60.
82. Borden RC, Bedient PB. Transport of dissolved hydrocarbons influenced by oxygen-limited biodegradation: 1. Theoretical development. Water Resour Res. 1986;22(13):1973–82.
83. Hers I, Atwater J, Li L, Zapf-Gilje R. Evaluation of vadose zone biodegradation of BTX vapours. J Contam Hydrol. 2000;46(3–4):233–64.
84. Talimcioglu NM. A model for evaluation of the impact of contaminated soil on groundwater. Ph.D. Dissertation,. Hoboken, NJ: Stevens Institute of Technology; 1991.
85. Korfiatis GP, Talimcioglu NM. Model for evaluation of the impact of contaminated soil on groundwater. Hoboken, NJ: Stevens Institute of Technology; 1991.
86. Korfiatis GP, Talimcioglu NM. IMPACT: a model for calculation of soil cleanup levels. Remediation. 1994;4(2):175–88.
87. Abreu LDV, Johnson PC. Simulating the effect of aerobic biodegradation on soil vapor intrusion into buildings: Influence of degradation rate, source concentration, and depth. Environ Sci Technol. 2006;40(7):2304–15.
88. Abreu LDV, Ettinger R, McAlary T. Simulated soil vapor intrusion attenuation factors including biodegradation for petroleum hydrocarbons. Ground Water Monit Remediat. 2009;29(1):105–17.
89. Yao YJ. Modeling vapor intrusion - the influence of biodegradation and useful approximation techniques. Ph.D. Dissertation. Providence, RI: Brown University; 2012.
90. Wiedemeier TH, Wilson JT, Kampbell DH, Miller RN, Hansen JE. Technical protocol for implementing intrinsic remediation with long-term monitoring for natural attenuation of fuel contamination dissolved in groundwater, vol. 1. Air Force Center for Environmental Excellence, Technology Transfer Division Brooks AFB: San Antonio, TX; 1995.
91. Ribbons DW, Eaton RW. Chemical transformations of aromatic hydrocarbons that support the growth of microorganisms. In: Chakrabarty AM, editor. Biodegradation and Detoxification of Environmental Pollutants. Boca Raton: CRC Press; 1982.
92. API. BioVapor Indoor Vapor Intrusion Model. 2021., https://www.api.org/oil-and-natural-gas/environment/clean-water/ground-water/vapor-intrusion/biovapor.

93. Verginelli I, Baciocchi R. Vapor intrusion screening model for the evaluation of risk-based vertical exclusion distances at petroleum contaminated sites. Environ Sci Technol. 2014;48 (22):13263–72.
94. Patterson BM, Davis GB. Quantification of vapor intrusion pathways into a slab-on-ground building under varying environmental conditions. Environ Sci Technol. 2009;43(3):650–6.
95. Verginelli I, Yao YJ, Suuberg EM. An excel®-based visualization tool of two-dimensional soil gas concentration profiles in petroleum vapor intrusion. Ground Water Monit Remediat. 2016;36(2):94–100.
96. Guo YM, Holton C, Luo H, Dahlen P, Gorder K, Dettenmaier E, Johnson PC. Identification of alternative vapor intrusion pathways using controlled pressure testing, soil gas monitoring, and screening model calculations. Environ Sci Technol. 2015;49(22):13472–82.
97. Ferguson CC, Krylov VV, McGrath PT. Contamination of indoor air by toxic soil vapours: a screening risk assessment model. Build Environ. 1995;30(3):375–83.
98. Nazaroff WW, Lewis SR, Doyle SM, Moed BA, Nero AV. Experiments on pollutant transport from soil into residential basements by pressure-driven airflow. Environ Sci Technol. 1987;21 (5):459–66.
99. Dawson H, Kapuscinski R, Schuver H. EPA's vapor intrusion database: evaluation and characterization of attenuation factors for chlorinated volatile organic compounds and residential buildings. EPA 530-R-10-002. Washington, DC 20460: Office of Solid Waste and Emergency Response (OSWER); 2012.
100. Lahvis MA, Ettinger RA. Improving risk-based screening at vapor intrusion sites in California. Ground Water Monit Remediat. 2021;41(2):73–86.
101. Krylov VV, Ferguson CC. Contamination of indoor air by toxic soil vapours: the effects of subfloor ventilation and other protective measures. Build Environ. 1998;33(6):331–47.
102. Abreu LDV. A transient three dimensional numerical model to simulate vapor intrusion into buildings. Ph.D. Dissertation,. Tempe, AZ: Arizona State University; 2005.
103. Yu S, Unger AJA, Parker B. Simulating the fate and transport of TCE from groundwater to indoor air. J Contam Hydrol. 2009;107(3–4):140–61.
104. Olson DA, Corsi RL. Characterizing exposure to chemicals from soil vapor intrusion using a two-compartment model. Atmos Environ. 2001;35(24):4201–9.
105. Olson DA, Corsi RL. Fate and transport of contaminants in indoor air. Soil Sediment Contam. 2002;11(4):583–601.
106. Robinson AL, Sextro RG. Radon entry into buildings driven by atmospheric pressure fluctuations. Environ Sci Technol. 1997;31(6):1742–8.
107. Abreu LDV, Schuver H. Conceptual model scenarios for the vapor intrusion pathway. Washington, DC: Office of Solid Waste and Emergency Response; 2012.
108. Hulot C, Hazebrouck B, Gay G, Malherbe L, Pokryszka Z. Vapor emissions from contaminated soils into buildings: comparison between predictions from transport model and field measurements. In: International FZK/TNO conference on contaminated soil; 2003. pp 353–361.
109. Hers I, Zapf-Gilje R, Evans D, Li L. Comparison, validation, and use of models for predicting indoor air quality from soil and groundwater contamination. Soil Sediment Contam. 2002;11 (4):491–527.
110. Provoost J, Reijnders L, Swartjes F, Bronders J, Seuntjens P, Lijzen J. Accuracy of seven vapour intrusion algorithms for VOC in groundwater. J Soils Sediments. 2009;9(1):62–73.
111. Yao YJ, Wang Y, Verginelli I, Suuberg EM, Ye JF. Comparison between PVI2D and Abreu-Johnson's Model for Petroleum Vapor Intrusion Assessment. Vadose Zone J. 2016;15(11):11.
112. Holton C, Luo H, Dahlen P, Gorder K, Dettenmaier E, Johnson PC. Temporal variability of indoor air concentrations under natural conditions in a house overlying a dilute chlorinated solvent groundwater plume. Environ Sci Technol. 2013;47(23):13347–54.

Numerical Models of Vapor Intrusion 2

Numerical models are those solved without mathematical solutions in analytical form. Instead, the solutions of models are calculated with numerical methods such as finite difference or finite element to obtain an approximate solution [1]. Compared to the analytical VI models, most of which are one-dimensional (1D), numerical VI models are capable of simulating complicated soil vapor transport in two or three dimensions [2]. Thus, numerical models are often used to investigate the influences of key environmental factors, such as source depth and strength, in typical or non-typical VI scenarios. However, numerical models cannot be implemented in Excel spreadsheets and usually require relevant knowledge involving fluid dynamics and chemical transport. Not to mention the extra training to use modeling software. Those requirements make it challenging to be used by ordinary investigators. In fact, most users of numerical models are experienced, academic researchers. As a result, it is almost impossible for numerical models to be employed as VI risk screening tools.

Despite those differences between the numerical and analytical VI models, both follow similar ways to simulate the VI processes, release from the subsurface source, soil vapor transport, entry into the building, and the calculation of the indoor air concentration. Especially for the last, contaminant indoor air concentrations are usually estimated in an analytical equation, as described in the first chapter, since most numerical simulations only involve subsurface processes, i.e., soil gas transport.

2.1 The Soil Gas Transport Involving No Biodegradation

Like analytical VI models or VI screening models, the numerical models also share the same governing equation of soil gas transport but have different simplifications for specified scenarios. The general governing equation was already illustrated in the first chapter and would not be repeated here. Tables 2.1 and 2.2 summarize simplifications employed in the numerical models involving no biodegradations

Y. Yao, Q. Chen, *Vapor Intrusion Simulations and Risk Assessments*,
https://doi.org/10.1007/978-981-19-2700-3_2

Tables 2.1 The summary of numerical simulations without biodegradation

Main governing equation	Examples of model use
Transient diffusion $\phi_{g,w,s}\frac{\partial c_{ig}}{\partial t} = \nabla \cdot (D_i \nabla c_{ig})$ (2.1)	Guo et al. [3] [1]
Steady-state advection and diffusion $\nabla \cdot (q_g c_{ig}) = \nabla \cdot (D_i \nabla c_{ig})$ (2.2)	The Brown model (Bozkurt [4] [3], Bozkurt et al. [5] [3], Pennell et al. [6] [3], Yao et al. [7–10]) The ASU model (Abreu [11] [3], Abreu and Johnson [12] [3], Luo [13] [3], Abreu et al. [14] [3])
Transient advection and diffusion $\phi_{g,w,s}\frac{\partial c_{ig}}{\partial t} = -\nabla \cdot \left(q_g c_{ig}\right) + \nabla \cdot \left(D_i \nabla c_{ig}\right)$ (2.3)	The Brown mode (Yao et al. [10, 15, 16] [3], Man et al. [17] [3]) The ASU model (Abreu [11] [3], Luo [13] [3], Abreu et al. [14] [3]) Qi et al. [18] [1]

Note: while some models were developed in more complete form, what is shown here is how the models were actually used in the indicated references. [1(3)] means 1(3)D models with numerical solutions. The definitions of all symbols are the same as those in Chap. 1
Adapted with permission from [1]. Copyright 2013 American Chemical Society

Table 2.2 The summary of numerical simulations involving biodegradation

Governing equation	Models
Steady diffusion and biodegradation (2 species) $0 = \nabla \cdot (D_i \nabla c_{ig}) - R_i$ (2.4)	Chen et al. [19] [1] The Brown model (Yao et al. [20, 21] [1])
Steady advection and diffusion and biodegradation (2 species) $0 = -\nabla \cdot (q_g c_{ig}) + \nabla \cdot (D_i \nabla c_{ig}) - R_i$ (2.5)	R-UNSAT [22] [2] VADBIO [23] [2] IMPACT [24–26] [2] MIN3P-DUSTY (Hers et al. [27]) [2] The Brown model (Yao [28] [3], Yao et al. [20, 29] [3], Yao et al. [21] [2]) The ASU model (Luo [30] [3], Abreu et al. [14, 31] [3] Ma et al. [32, 33] [3]) CompFlow Bio [34] [2]
Transient advection and diffusion and biodegradation (2 species) $\phi_{g,w,s}\frac{\partial c_{ig}}{\partial t} = -\nabla \cdot \left(q_g c_{ig}\right) + \nabla \cdot \left(D_i \nabla c_{ig}\right) - R_i$ (2.6)	Picone et al. [35] (STOMP) [1] R-UNSAT [22] [2] VADBIO [23] [2] IMPACT [24–26] [2] The Brown model (Liu et al. [36] [2]) The ASU model(Luo [13] [3], Abreu et al. [14, 31] [3]) CompFlow Bio [34] [2]

Note: [1(2, 3)] means 1(2, 3)D models. The definitions of all symbols are the same as those in Chap. 1
Adapted with permission from [1]. Copyright 2013 American Chemical Society

and those involving biodegradation, respectively. It should be noted that those involving a preferential pathway will be discussed in Chap. 5.

As discussed in the first chapter, the intrinsic characteristics of 1D models limit their application when facing heterogeneous upper boundary conditions. However, if the focus was shifted from the prediction of indoor air concentration to the estimation of upward soil gas flux of contaminants, it is reasonable to assume a uniform open ground surface as the upper boundary condition by employing only 1D simulations with the consideration of lateral transport.

For example, Guo et al. [3] employed the software of HYDRUS 1D to develop a 1D numerical model to study the temporal variations of the upward flux in the unsaturated zone due to the fluctuation of the water table. In their simulated scenario, the soil surface was assumed to open to the atmosphere. They found that the time-averaged surface of vapor fluxes for cyclic water table elevations was greater than for stationary water table conditions at an equivalent time-averaged water table position (flux changes were generally less than 50%). Simulation results also suggested that emission flux changes due to groundwater fluctuation are likely to be significant at sites with shallow groundwater (e.g., <0.5 m) and permeable soil types (e.g., sand). Qi et al. [18] developed a 1D numerical model to simulate VOC mass transfer from groundwater to the atmosphere. They found that groundwater table fluctuations can increase VOC fluxes from coarse or medium sand soils, while the flux is not significantly affected in silt or clay soils.

In other cases, numerical models are often used for more complicated scenarios, considering both horizontal and vertical transport. There are currently two numerical VI models mostly used. One was developed by Abreu and Johnson from Arizona State University [12], while the other was first introduced by Bozkurt et al. from Brown University in 2009 [4]. The first was coded by Abreu in 2005 [11] and solved with finite difference method, and the other was introduced based on a commercial software package named Comsol Multiphysics and solved with the finite element method [5, 6]. As those two models were further modified and used by a few different users [7, 13], they were named the ASU model and the Brown model, respectively, in this book.

The ASU model was first introduced by Abreu and Johnson [12] to simulate the soil vapor-to-indoor air pathway. The model was first used to simulate typical VI scenarios described in work by Johnson and Ettinger [37], involving a homogenous infinite source, at least over the model domain. The perimeter crack in the building foundation is assumed to be the migration pathway of soil vapor advection caused by the pressure difference between indoor air and the surface air at the open ground surface surrounding the building of concern. These simulation results were later compared to Yao et al.'s analytical solution [38], confirming that the developed analytical solution can predict accurate subslab perimeter crack concentrations as numerical simulations. In Abreu and Johnson's study [12], the ASU model was also applied to simulate the scenario described in the work by Lowell and Eklund [39], examining the lateral separation between the vapor source and the building of concern. In numerical simulations, the influences of the building foundation and advective soil flow caused by pressure differences were included. The results are still in good agreement with Lowell and Eklund's analytical solutions [39]. After that, the ASU model was modified to simulate the wind effect on soil gas concentration profiles and indoor air concentrations based on scenarios involving different wind directions and strengths in Luo's dissertation [13]. The ASU model was also employed in an EPA technical document by Abreu et al. [14] in 2012 to simulate conceptual model scenarios of VI, involving transient behaviors of contaminant soil gas with heterogeneous soil properties and building construction arrangements.

The Brown model was first introduced by Bozkurt et al. [4, 5] and Pennell et al. [6] to simulate various scenarios involving heterogeneous soil properties and different arrangements of surface buildings and pavements [6, 7, 11, 30]. These scenarios are usually too sophisticated to be solved with analytical solutions. In most simulations, the modeling parameters were similar to Abreu and Johnson [12], except for the boundary condition at the foundation slab crack. In the module of Darcy's law, a pressure boundary condition and a flux boundary condition were applied at the crack in the Brown model and the ASU model, respectively [4, 5, 12]. The Brown model was later employed by Yao et al. [7] to compare with the J-E model in typical VI scenarios. The results indicate that the J-E model may provide an over-estimate compared to the numerical simulations with soil permeability lower than 10^{-11} m^2. In another study by Yao et al. [8], the Brown model was used to examine the influences of the crack locations in determining indoor air concentration, and it was reported that the shape and the location of the slab crack do not affect the contaminant mass entry rate into the building significantly, and only the size of cracks matters.

After that, the Brown model was applied to examine the lateral transport of soil gas in cases involving layering and impermeable surface covering, and the simulated results were compared with an analytical model [40]. Both predictions suggest that buildings greater than 30 m from a contaminant plume boundary can still be affected by VI with any two of the three factors, strong source, shallow source depth, and significant surface cover. This finding justifies the concern of the US EPA about applying the 30 m lateral separation distance for risk screens in the presence of physical barriers (e.g., asphalt covers or ice) at the ground surface [41]. It is worth noting that more than 70% recorded lateral separation distances between buildings and groundwater monitoring wells, where the sample was taken and measured as the corresponding source concentrations, are higher than 30 m [42].

In 2017, the Brown model was again used to investigate the role of soil texture in VI involving a groundwater source. The simulated 1D profiles of soil gas concentration were first validated using results of soil column experiments, and then the 3D simulated results were compared with statistical analysis of USEPA's VI database [43]. Consistent with the latter, the 3D simulations indicate that soil particle texture can play a role in determining subslab-to-indoor air concentration attenuation. There is no apparent relationship between soil particle size and groundwater source-to-subslab except in the case of a shallow contaminant source. They found that, although soil particle texture can play a role in determining subslab-to-indoor air concentration attenuation, there is no apparent relationship between soil particle size and groundwater source-to-subslab attenuation factor except in the case of a shallow contaminant source.

These findings were further justified by Yao et al. [9] in examining VI involving a two-layer system by employing the Brown model. The simulated results show that if the subfoundation soil permeability is larger than 10^{-11} m^2, the advection dominates the soil gas transport into the building. The indoor air concentration increases by half an order of magnitude with one order of magnitude increase in the subfoundation soil permeability. Otherwise, diffusion plays a more critical role and the subfoundation soil texture does not cause significant variation in indoor air concentration. The same study also found that the major resistance to the upward diffusion from a groundwater source is actually determined by the thickness of the capillary fringe, which is dependent on soil texture.

In a recent study by Man et al. [17], the Brown model was adapted with a floating boundary condition to simulate the water table fluctuations, the influences of which were examined in VI scenarios involving a groundwater source. The modified model was first validated with measured results from a sandbox experiment and later used in scenarios involving different depths, periods, and amplitudes of groundwater level. It was reported that soil texture plays a vital role in determining the impact of the fluctuations, especially in cases with soil particle sizes between 0.25 and 0.44 mm.

2.2 The Soil Gas Transport Involving Biodegradation

In the current literature, the ASU model and the Brown model are mostly used for numerical simulations to include biodegradation in petroleum vapor intrusion (PVI).

After the initial introduction [12], the ASU model was further modified by including additional biodegradation and oxygen transport in Abreu and Johnson [44]. The simulations are based on the assumption that oxygen-limited biodegradation only occurs in the presence of oxygen concentration higher than 5% of the atmospheric concentration. Based on the coupled transport and reaction, the depths of the aerobic/anaerobic interface simulated by the 3D numerical model fit the analytical predictions by Roggemans et al.'s hypothesis [45] when the location of the interface is far from the building. The ASU model was also employed by Abreu et al. [31] to provide estimates of indoor air concentrations based on numerical simulations of typical VI scenarios, the same as those in Abreu and Johnson [44] except for different parameters. With the help of the ASU model, an EPA technical document by Abreu et al. [14] was later released to simulate the role of oxygen-limited biodegradation in various conceptual scenarios.

A sensitivity test of the ASU model was performed for petroleum hydrocarbons by Ma et al. [46] in 2016, which reported that the uncertainties in scenarios involving strong and shallow vapor sources are much less than those in cases with low-concentration and deep sources. In the former cases, influences of oxygen-limited biodegradation in the soil gas concentration attenuation are limited since the oxygen may not be available in the subfoundation zone. The ASU model was also used in Ma et al. [32] to estimate the threat to human health and safety in buildings overlaying ethanol-blended fuel impacted sites. In their simulations, advection was included to evaluate VI risks of methane and benzene. The study concluded that if methanogenic activity near the source zone is sufficiently high to cause advective gas transport, the indoor methane concentration may exceed the flammable threshold under simulated conditions. In another study by Ma et al. [33], the ASU model was employed to investigate the VI risks of fuel ether oxygenates methyl tert-butyl ether (MTBE), tert-amyl methyl ether (TAME), and ethyl tert-butyl ether (ETBE). They reported that MTBE is more dangerous than the other two in causing PVI problems and claimed that the vertical screening distances recommended by USEPA might not be enough to prevent PVI risks by fuel additives, especially ether oxygenates.

Compared to the piecewise first-order biodegradation in Abreu and Johnson [44] and Abreu et al. [31], in which the reaction rate is a function of contaminant concentration in the aerobic zone, in Yao's dissertation [28], second-order biodegradation involving both oxygen and contaminant concentration was simulated with the Brown model based on incorporating biodegradation kinetics and oxygen transport into earlier developed model structures [4–7, 10, 38, 47, 48]. Those results were also published in Yao et al. [29] and indicated that the aerobic/anoxic interface depth is determined by the ratio of contaminant source vapor to atmospheric oxygen concentration. The contaminant concentration profile in the aerobic zone was significantly influenced by the choice of rate law.

Yao et al. [20] used the Brown model to examine the role of soil texture in PVI in comparison with a 1D analytical solution. The comparison showed that the most significant attenuation of petroleum products in the vertical transport could be expected in the fine-grain soil. Based on these findings, the authors proposed 3 and 5 m s soil-type-dependent vertical screening distances for fine and coarse-grained soils, respectively. Later, the same model was used to investigate the detailed effects of the capillary fringe on petroleum vapor biodegradation and attenuation [21]. The authors found that it is highly unlikely for groundwater sources to induce unacceptable petroleum vapor intrusion risks. On the other hand, the simulations suggested that the vertical smear zone of residual LNAPL (light non-aqueous phase liquid) contamination, induced by the water table fluctuations, may lead to a potential VI threat if with a short vertical source-building separation distance, and thus requires more attention. Liu et al. [36] employed the Brown model to simulate the temporal trend of subfoundation of soil gas concentration profiles after injecting fresh air into the subsurface. They reported that the microbe activities of consuming petroleum soil vapors would deplete the oxygen in the subfoundation in several weeks to 1 month at a specific site described in Luo et al. [13]

In addition to the applications of the ASU and Brown models, a few studies employed other numerical models to investigate the influences of environmental factors in PVI. For instance, a numerical model of R-UNSAT [22] was used to study two-dimensional (2D) transport of multiple species in scenarios with radially symmetric boundaries. However, the model can only simulate a constant advection in the water phase, and it is impossible to include the construction features of the building foundation. Another example is the IMPACT model, which was developed initially by Talimcioglu in 1991 [24–26] to provide soil cleanup criteria for hazardous waste sites, while in Sanders and Talimcioglu's study [49], the model was applied to estimate the volatilization of VOCs. The IMPACT model is similar to Jury's pesticide model, except the former includes an additional hydrodynamic dispersion, which is helpful in the examination of the seasonal effects and short-term influences of temperature change and rainfall events. Hers et al. [23] used VADBIO to study the building foundation floor in the coupled transport of oxygen and contaminants with biodegradation, which was simulated in different reaction kinetics. Yu et al. [34] employed a multiphase compositional model of CompFlow to investigate soil gas transport involving a groundwater source in a scenario defined according to a field site. In their study, the difference from the usual VI modeling is that the transport of dense nonaqueous phase liquid (DNAPL) contaminant into the groundwater was also included. Picone et al. [35] used the code of "Subsurface Transport Over Multiple Phases" (STOMP) to simulate vertical moisture content profiles and study the role of aerobic biodegradation in VI risk assessments. The 1D numerical results showed that up to two orders of magnitude difference in soil gas concentration attenuation factor could be induced by the enhanced diffusion due to the water table fluctuation. Chen et al. [19] developed a 1D numerical model to study the influences of the mass transfer among the solid, water, and gas phases in determining the first-order biodegradation rate in the vadose zone. In their simulations, a steady-state soil moisture distribution in the unsaturated zone and first-order kinetic models to describe the interphase mass transfer. They claimed that the biodegradation rate could be overestimated if the reaction kinetics was simplified as an instantaneous reaction thus under-predicting upward vapor fluxes at the subfoundation.

2.3 Soil Gas Entry into the Building

Most numerical simulations shared the same assumptions and even governing equations to estimate the contaminant soil gas entry rate into the building, as already discussed in Chap. 1. The difference lies in the way of obtaining subslab soil gas concentration and volumetric flow rate into the building, which can be either numerically simulated or analytically calculated. Thus, there has been relatively less work for soil gas entry into the buildings based on numerical models than the number of studies involving soil gas transport.

Song et al. [50, 51] introduced some mathematical equations to describe the impacts of environmental factors, such as indoor pressure conditions, wind effect, and crack size, on the volumetric soil gas entry rate and indoor-outdoor air exchange rate, between which a positive correlation was reported. According to their results, these two factors dominate the subslab-to-indoor air contaminant concentration attenuation, and the influence of indoor pressure or wind effects is quite limited, against the conclusion of numerical simulations by Luo [13], which claimed the direction and strength of wind could play a significant role in determining contaminant indoor air concentration.

To decrease the uncertainty in estimating indoor air concentration, Diallo et al. [52] proposed a 2D semi-empirical model to predict the contaminant entry rate in the building for both volatile organic compounds and radon. Their model was applied in scenarios involving different building foundations, including crawl space, supported slab, and floating slab. Later, the numerical method was further simplified with an airflow analytical model previously introduced by the same authors in 2013 [53] to quantify the influences of the subfoundation porous layer on the volumetric soil gas entry rate into the building structure [54].

In some cases, the temperature was also believed to play a role in the entry process. Some empirical equations to estimate indoor air concentrations were developed by Barnes and McRae [55] by including the influences of both soil temperature and atmospheric pressure conditions. They claimed that a five-degree increase in soil temperature could cause the increase of contaminant indoor concentration by a factor of two.

To provide an independent line of evidence to augment empirical attenuation factors, McAlary et al. [56] presented a fluid model to estimate the subslab-to-indoor air soil gas concentration attenuation factor by including the influences of the subslab pressure and flow, as well as building operational conditions. Later, to assess the radius of influence for subslab venting systems, McAlary et al. [57] developed a mathematical model representing a two-layer system with horizontal radial flow through transmissive material below the floor slab and vertical flow through discontinuities in the floor slab.

In a classical VI scenario, the soil gas enters the building of concern by advection induced by the indoor-outdoor pressure difference. Based on this assumption, it requires the building should be surrounded by an open ground surface. However, it is not always the case, especially in urban areas where the ground surface is often paved with asphalt or concrete. Those surface pavements, which are usually considered of lower permeability than soil, are supposed to block the surface air into the soil to cause the soil gas advection based on the classic assumption thus decreasing VI risks in those cases. However, there have not been any such results reported by past literature. To examine the VI threat in scenarios involving a building surrounded by a paved ground surface, Yao et al. [16] used the Brown model to simulate the influences of fluctuated indoor pressure in inducing upward soil gas advective flow, which could help draw subfoundation contaminants into the building, as shown in Fig. 2.1. The authors found that the high-frequency fluctuation of indoor pressure can play an important role, especially in cases with surface pavements causing bidirectional advective soil gas flow due to the blocking effect of the surface pavement. As a result, the VI potential for buildings surrounded by paved ground surface can be higher than that in cases with an open ground surface.

2.4 Calculation of Indoor Air Concentration

As discussed in Chap. 1, the most common assumption used to calculate indoor air concentration is to consider the enclosed space of the building as one or multiple continuous stirred tanks. This assumption works on preconditions by considering the temporal and spatial variabilities of indoor air concentrations negligible. It requires a mass balance between the contaminant intrusion rate into the building and the exchange rate from indoor to outdoor. Such simplification is reasonable in risk screens for estimating a conservative or averaged exposure in the long term.

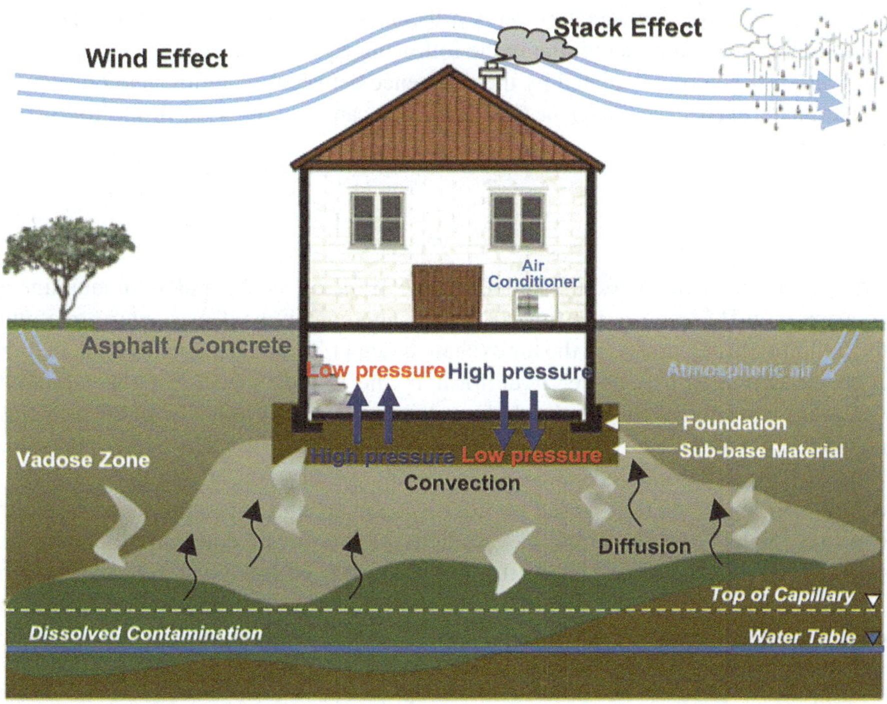

Fig. 2.1 Soil gas advection into the building by indoor pressure fluctuation. Represented with permission from [16]. Copyright 2019 Elsevier B.V

However, when it is necessary to predict the temporal behaviors of indoor air concentrations in different building compartments, this assumption may no longer be valid.

Shirazi and Pennell [58] and later Shirazi et al. [59] introduced a modeling system that employed the Brown model to simulate subsurface soil gas transport and CFD0 (i.e., Computational Fluid Dynamics incorporated in CONTAM) model to predict surface airflow surrounding the building of concern. Then the results of the two models were used as the input of the CONTAM model to predict the indoor concentration profiles. Compared to the standard VI modeling, the new system extended the common subsurface fate and transport equations to incorporate wind and stack effects on indoor air pressure, building air exchange rate (AER), and indoor contaminant concentration to improve VI exposure risk estimates. Similar to Luo's points [30], their studies also concluded that the wind speed and strength, as well as the opening locations, should not be ignored in determining building conditions and thus indoor air contaminant concentration.

In order to simulate the temporal behavior of indoor air concentration, Ström et al. [60] introduced an equation called Unsteady-CST to calculate the transient indoor air concentration. In studied scenarios, the contaminant mass entry rate is no longer equal to the exchange rate of contaminant from indoor air to outdoor air, but the contaminant indoor air concentration is still considered homogenous in the enclosed space of the indoor air. According to the following equation, the change in indoor air concentration can be calculated by the difference between the inflow and outflow of contaminant mass. The transient CST equation can be expressed in the following form [60]

$$V\frac{dc_{\mathrm{in}}}{dt} = J_{\mathrm{ck}} - c_{\mathrm{in}}A_{\mathrm{e}}V \tag{2.7}$$

where V is the volume of the indoor space [L^3], c_{in} is the indoor contaminant concentration [M/L^3], t is the time [T], J_{ck} is the contaminant mass entry rate into the building [M/T], and A_{e} is the air exchange rate [1/T]. In this equation, the indoor concentration is still considered uniform in the indoor space, but the temporal variations were considered. It should be noted that this is different from the previous assumption as employed in previous numerical studies [14], where the indoor air concentration is considered linear to contaminant entry rate all the time if with a constant air exchange rate. Based on Eq. (2.7), the change in contaminant entry rate can immediately induce temporal variations of indoor air concentration, but the impact will not be significant in the initial period, which could be hours. Such difference is important for modeling involving short-term temporal variations of contaminant indoor air concentration, e.g., for cases involving the manipulation of indoor pressure to identify the potential presence of a preferential pathway.

References

1. Yao YJ, Shen R, Pennell KG, Suuberg EM. A review of vapor intrusion models. Environ Sci Technol. 2013;47(6):2457–70.
2. Verginelli I, Yao YJ. A review of recent vapor intrusion modeling work. Ground Water Monit Remediat. 2021;41(2):138–44.
3. Guo YM, Holton C, Luo H, Dahlen P, Johnson PC. Influence of fluctuating groundwater table on volatile organic chemical emission flux at a dissolved chlorinated-solvent plume site. Ground Water Monit Remediat. 2019;39(2):43–52.
4. Bozkurt O. Investigation of vapor intrusion scenarios using a three-dimensional numerical model. Ph.D. Dissertation,. Providence, RI: Brown University; 2009.
5. Bozkurt O, Pennell KG, Suuberg EM. Simulation of the vapor intrusion process for nonhomogeneous soils using a three-dimensional numerical model. Ground Water Monit Remediat. 2009;29(1):92–104.
6. Pennell KG, Bozkurt O, Suuberg EM. Development and application of a three-dimensional finite element vapor intrusion model. J Air Waste Manage Assoc. 2009;59(4):447–60.
7. Yao YJ, Shen R, Pennell KG, Suuberg EM. Comparison of the Johnson-Ettinger vapor intrusion screening model predictions with full three-dimensional model results. Environ Sci Technol. 2011;45(6):2227–35.
8. Yao YJ, Pennell KG, Suuberg EM. Simulating the effect of slab features on vapor intrusion of crack entry. Build Environ. 2013;59:417–25.
9. Yao YJ, Xiao YT, Mao F, Chen HY, Verginelli I. Examining the role of sub-foundation soil texture in chlorinated vapor intrusion from groundwater sources with a two-layer numerical model. J Hazard Mater. 2018;359:544–53.

10. Yao YJ, Pennell KG, Suuberg E. Vapor intrusion in urban settings: Effect of foundation features and source location. In: Conference on urban environmental pollution - overcoming obstacles to sustainability and quality of life, Elsevier Science Bv: Boston, MA; 2010. p. 245.
11. Abreu LDV. A transient three dimensional numerical model to simulate vapor intrusion into buildings. Ph.D. Dissertation,. Tempe, AZ: Arizona State University; 2005.
12. Abreu LDV, Johnson PC. Effect of vapor source - building separation and building construction on soil vapor intrusion as studied with a three-dimensional numerical model. Environ Sci Technol. 2005;39(12):4550–61.
13. Luo H. Field and modeling studies of soil vapor migration into buildings at petroleum hydrocarbon impacted sites. . Ph.D. Dissertation,. Tempe, AZ: Arizona State University; 2009.
14. Abreu LDV. Conceptual model scenarios for the vapor intrusion pathway. Washington DC: Office of Solid Waste and Emergency Response; 2012.
15. Yao YJ, Mao F, Ma SS, Yao YH, Suuberg EM, Tang XJ. Three-dimensional simulation of land drains as a preferential pathway for vapor intrusion into buildings. J Environ Qual. 2017;46(6): 1424.
16. Yao YJ, Xiao YT, Luo J, Wang GF, Ström J, Suuberg E. High-frequency fluctuations of indoor pressure: a potential driving force for vapor intrusion in urban areas. Sci Total Environ. 2020;710:136309.
17. Man J, Wang GF, Chen Q, Yao YJ. Investigating the role of vadose zone breathing in vapor intrusion from contaminated groundwater. J Hazard Mater. 2021;416:11.
18. Qi SQ, Luo J, O'Connor D, Cao XY, Hou DY. Influence of groundwater table fluctuation on the non-equilibrium transport of volatile organic contaminants in the vadose zone. J Hydrol. 2020;580:124353.
19. Chen YM, Hou DY, Lu CH, Spain JC, Luo J. Effects of rate-limited mass transfer on modeling vapor intrusion with aerobic biodegradation. Environ Sci Technol. 2016;50(17):9400–6.
20. Yao YJ, Mao F, Xiao YT, Chen HY, Verginelli I, Luo J. Investigating the role of soil texture in petroleum vapor intrusion. J Environ Qual. 2018;47(5):1179–85.
21. Yao YJ, Mao F, Xiao YT, Luo J. Modeling capillary fringe effect on petroleum vapor intrusion from groundwater contamination. Water Res. 2019;150:111–9.
22. Lahvis MA, Baehr AL. Documentation of R-UNSAT, a computer model for the simulation of reactive, multispecies transport in the unsaturated zone. Reston, VA: US Department of the Interior, US Geological Survey; 1997.
23. Hers I, Atwater J, Li L, Zapf-Gilje R. Evaluation of vadose zone biodegradation of BTX vapours. J Contam Hydrol. 2000;46(3–4):233–64.
24. Talimcioglu NM. A model for evaluation of the impact of contaminated soil on groundwater. Ph.D. Dissertation,. Hoboken, NJ: Stevens Institute of Technology; 1991.
25. Korfiatis GP, Talimcioglu NM. Model for evaluation of the impact of contaminated soil on groundwater. Hoboken, NJ: Stevens Institute of Technology; 1991.
26. Korfiatis GP, Talimcioglu NM. IMPACT: a model for calculation of soil cleanup levels. Remediation. 1994;4(2):175–88.
27. Hers I, Jourabchi P, Lahvis MA, Dahlen P, Luo EH, Johnson P, DeVaull GE, Mayer KU. Evaluation of seasonal factors on petroleum hydrocarbon vapor biodegradation and intrusion potential in a cold climate. Ground Water Monit Remediat. 2014;34(4):60–78.
28. Yao YJ. Modeling vapor intrusion - the influence of biodegradation and useful approximation techniques. . Ph.D. Dissertation,. Providence, RI: Brown University; 2012.
29. Yao Y, Shen R, Pennel KG, Suuberg EM. A numerical investigation of oxygen concentration dependence on biodegradation rate laws in vapor intrusion. Environ Sci Process Impacts. 2013;15(12):2345–54.
30. Luo H, Dahlen P, Johnson PC, Peargin T, Creamer T. Spatial variability of soil-gas concentrations near and beneath a building overlying shallow petroleum hydrocarbon–impacted soils. Ground Water Monit Remediat. 2009;29(1):81–91.

31. Abreu LDV, Ettinger R, McAlary T. Simulated soil vapor intrusion attenuation factors including biodegradation for petroleum hydrocarbons. Ground Water Monit Remediat. 2009;29(1): 105–17.
32. Ma J, Luo H, DeVaull GE, Rixey WG, Alvarez PJ. Numerical model investigation for potential methane explosion and benzene vapor intrusion associated with high-ethanol blend releases. Environ Sci Technol. 2014;48(1):474–81.
33. Ma J, Xiong D, Li H, Ding Y, Xia X, Yang Y. Vapor intrusion risk of fuel ether oxygenates methyl tert-butyl ether (MTBE), tert-amyl methyl ether (TAME) and ethyl tert-butyl ether (ETBE): a modeling study. J Hazard Mater. 2017;332:10–8.
34. Yu S, Unger AJA, Parker B. Simulating the fate and transport of TCE from groundwater to indoor air. J Contam Hydrol. 2009;107(3–4):140–61.
35. Picone S, Valstar J, Van Gaans P, Grotenhuis T, Rijnaarts H. Sensitivity analysis on parameters and processes affecting vapor intrusion risk. Environ Toxicol Chem. 2012;31(5):1042–52.
36. Liu YQ, Man J, Wang Y, Xiao YT, Tang W, Chen Q, Yao Y. J., numerical study of the building pressure cycling method for evaluating vapor intrusion from groundwater contamination. Environ Sci Pollut Res. 2020;27(28):35416–27.
37. Johnson PC, Ettinger RA. Heuristic model for predicting the intrusion rate of contaminant vapors into buildings. Environ Sci Technol. 1991;25(8):1445–52.
38. Yao YJ, Pennell KG, Suuberg EM. Estimation of contaminant subslab concentration in vapor intrusion. J Hazard Mater. 2012;231:10–7.
39. Lowell PS, Eklund B. VOC emission fluxes as a function of lateral distance from the source. Environ Prog. 2004;23(1):52–8.
40. Yao YJ, Wu Y, Suuberg EM, Provoost J, Ma JQ, Liu J. Vapor intrusion attenuation factors relative to subslab and source, reconsidered in light of background data. J Hazard Mater. 2015;286:553–61.
41. US EPA. Draft guidance for evaluating the vapor intrusion to indoor air pathway from groundwater and soils (subsurface vapor intrusion guidance). Washington DC: Office of Solid Waste and Emergency Response (OSWER). EPA530-D-02-004.; 2002.
42. Yao YJ, Verginelli I, Suuberg EM, Eklund B. Examining the use of USEPA's generic attenuation factor in determining groundwater screening levels for vapor intrusion. Ground Water Monit Remediat. 2018;38(2):79–89.
43. Yao YJ, Wang Y, Zhong Z, Tang ML, Suuberg EM. Investigating the role of soil texture in vapor intrusion from groundwater sources. J Environ Qual. 2017;46(4):776.
44. Abreu LDV, Johnson PC. Simulating the effect of aerobic biodegradation on soil vapor intrusion into buildings: influence of degradation rate, source concentration, and depth. Environ Sci Technol. 2006;40(7):2304–15.
45. Roggemans S, Bruce CL, Johnson PC, Johnson RL. Vadose zone natural attenuation of hydrocarbon vapors: an emperical assessment of soil gas vertical profile data. Washington DC: American Petroleum Institute; 2001.
46. Ma J, Yan G, Li H, Guo S. Sensitivity and uncertainty analysis for Abreu & Johnson numerical vapor intrusion model. J Hazard Mater. 2016;304:522–31.
47. Yao YJ, Pennell KG, Suuberg EM. The influence of transient processes on vapor intrusion processes. In: Paper presented at the air and waste management association vapor intrusion conference, Chicago, IL; 2010. pp. 29–30.
48. Yao YJ, Shen R, Pennell KG, Suuberg EM. Estimation of contaminant subslab concentration in vapor intrusion including lateral source-building separation. Vadose Zone J. 2013;12(3):9.
49. Sanders PF, Talimcioglu NM. Soil-to-indoor air exposure models for volatile organic compounds: the effect of soil moisture. Environ Toxicol Chem. 1997;16(12):2597–604.
50. Song S, Schnorr BA, Ramacciotti FC. Quantifying the influence of stack and wind effects on vapor intrusion. Hum Ecol Risk Assess. 2014;20(5):1345–58.
51. Song S, Schnorr BA, Ramacciotti FC. Accounting for climate variability in vapor intrusion assessments. Hum Ecol Risk Assess. 2018;24(7):1838–51.

52. Diallo TM, Collignan B, Allard F. 2D semi-empirical models for predicting the entry of soil gas pollutants into buildings. Build Environ. 2015;85:1–16.
53. Diallo TMO, Collignan B, Allard F. Analytical quantification of airflows from soil through building substructures. Build Simul. 2013;6(1):81–94.
54. Diallo TMO, Collignan B, Allard F. Analytical quantification of the impact of sub-slab gravel layer on the airflow from soil into building substructures. Build Simul. 2018;11(1):155–63.
55. Barnes DL, McRae MF. The predictable influence of soil temperature and barometric pressure changes on vapor intrusion. Atmos Environ. 2017;150:15–23.
56. McAlary TA, Gallinatti J, Thrupp G, Wertz W, Mali D, Dawson H. Fluid flow model for predicting the intrusion rate of subsurface contaminant vapors into buildings. Environ Sci Technol. 2018;52(15):8438–45.
57. McAlary T, Wertz W, Mali D, Nicholson P. Mathematical analysis and flux-based radius of influence for radon/VOC vapor intrusion mitigation systems. Sci Total Environ. 2020;740: 139988.
58. Shirazi E, Pennell KG. Three-dimensional vapor intrusion modeling approach that combines wind and stack effects on indoor, atmospheric, and subsurface domains. Environ Sci Process Impacts. 2017;19(12):1594–607.
59. Shirazi E, Hawk GS, Holton CW, Stromberg AJ, Pennell KG. Comparison of modeled and measured indoor air trichloroethene (TCE) concentrations at a vapor intrusion site: influence of wind, temperature, and building characteristics. Environ Sci Process Impacts. 2020;22(3): 802–11.
60. Strom JGV, Guo YM, Yao YJ, Suuberg EM. Factors affecting temporal variations in vapor intrusion-induced indoor air contaminant concentrations. Build Environ. 2019;161:9.

US EPA's Vapor Intrusion Database and Generic Attenuation Factor

3

3.1 Introduction of US EPA's Vapor Intrusion Database

The US EPA's Vapor Intrusion (VI) database is a compilation of measurements from VI sites throughout the USA. The database contains information from 42 sites and 21 kinds of chemicals, such as chlorinated chemicals and petroleum products, the latter of which however only cover only 3% of the whole data sets [1]. Other contaminants with VI potential, such as mercury or semi-volatile organic chemicals (SVOCs) are not involved [1]. In the database, 2929 total residence and chemical combinations are included. Of these measurements, "*1,021 (35 percent) are paired groundwater and indoor air measurements, 235 (8 percent) are paired exterior soil gas and indoor air measurements, 1,582 (54 percent) are paired subslab soil gas and indoor air measurements, and 91 (3 percent) are paired crawlspace and indoor air measurements*" [1]. The building types include "*residential (85 percent), institutional or commercial (10 percent), and multi-use (residential and non-residential) buildings (5 percent)*" [1].

Most of the datasets in EPA's database came from consultants and state regulators, but the EPA's regional offices also made some contributions. Information about both sampling and analytical methods was also collected and evaluated by US EPA to ensure the quality of measurements [2]. The database can be accessed with a spreadsheet compiling all measurements and some statistical analysis. Readers interested are directed to the EPA's VI website for downloading [3].

The primary objective of the setup of this database was to understand the soil gas concentration attenuation process from the subsurface to the indoor air and the generation of the generic attenuation factors of vapor concentration. As introduced in Chap. 1, there are two soil gas transport processes where the attenuation could occur in a typical VI scenario, soil gas transport from the vapor source to the subslab and from the subslab to the indoor air. It is believed that the two steps are independent of each other, which is the key to understanding the observations recorded in the database [4].

Y. Yao, Q. Chen, *Vapor Intrusion Simulations and Risk Assessments*,
https://doi.org/10.1007/978-981-19-2700-3_3

3.2 Examinations of the Database Based on Modeling

Figure 3.1 shows the dependence of measured groundwater source-to-indoor air concentration attenuation factor (c_{in}/c_s) on groundwater source vapor concentration (c_s), calculated based on Henry's law and measured groundwater concentrations recorded in the database. Figure 3.1a shows a trend that the attenuation factor of c_{in}/c_s decreases with the increases in c_s, against the common assumption employed by all VI modeling that the former should be independent of the latter in a steady-state without biodegradation. Figure 3.1b shows that such a trend still exists in the data with only chlorinated chemicals, which are considered difficult to biodegradable, leaving out the petroleum hydrocarbons, which are believed to be biodegradable in aerobic conditions [4].

This issue was further examined in Fig. 3.2 by plotting the measured subslab-to-indoor air concentration attenuation factor (c_{in}/c_{ss}) as a function of measured subslab soil vapor concentration c_{ss} based on the same database. Again, a similar trend was observed, showing the measured c_{in}/c_{ss} is inversely related to measured c_{ss}, as expected. Since the similar trends in two types of attenuation factors, which share the same numerator, there might be some problem with the measured c_{in} [4].

The unusual phenomenon in Fig. 3.2 can be explained by considering rough limits on absolute contaminant indoor air concentrations, such that most of c_{in} falls in a range between 0.1 and 10 μg/m^3. Thus, the application by using such a constant to be divided by c_{ss} or c_s would inevitably generate an inverting trend with the dominator. This consideration was further justified by the results shown in Fig. 3.3a, which plots c_{in} as a function of c_{ss}. As expected, there is no significant trend of c_{in} with c_{ss} for the majority of the data, and only if c_{ss} is higher than a

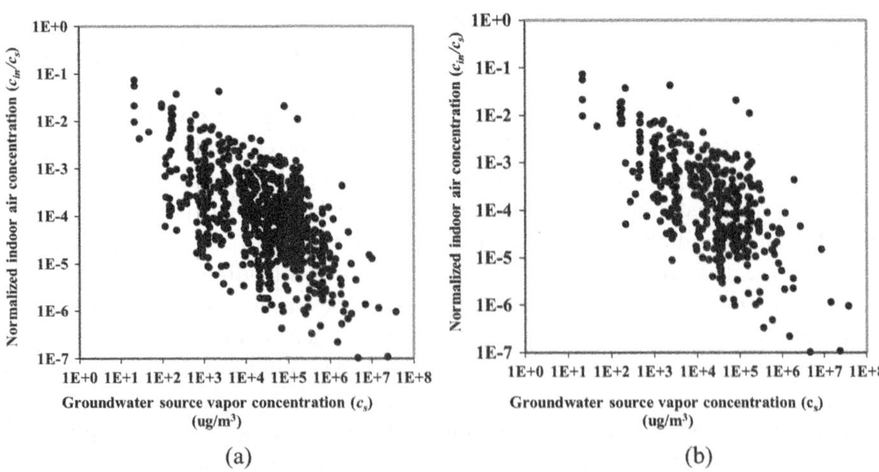

Fig. 3.1 Calculated groundwater source vapor-to-measured indoor air concentration attenuation factor for (**a**) all contaminants (**b**) Tetrachloroethylene (PCE) and Trichloroethylene (TCE) data taken from the US EPA's VI database [6]. Reprinted with permission from [4]. Copyright 2013 American Chemical Society

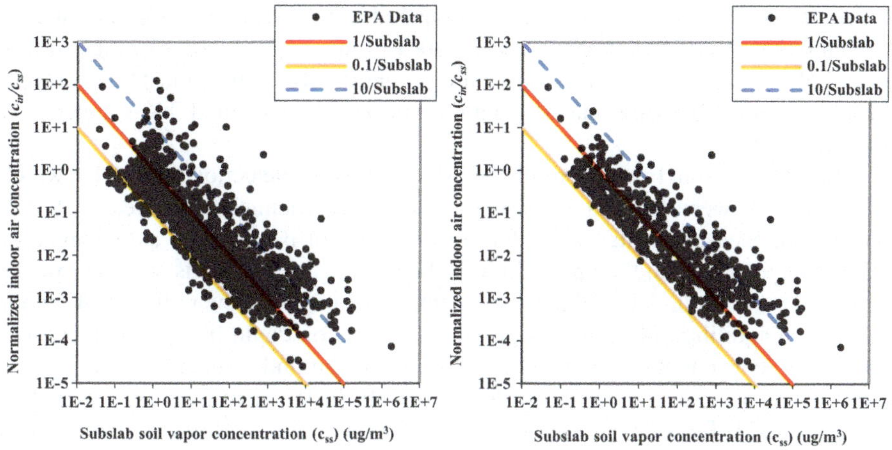

(a) (b)

Fig. 3.2 Measured subslab-to-indoor air concentration attenuation factor for (**a**) all contaminants (**b**) PCE and TCE in EPA's VI database [6]. Reprinted with permission from [4]. Copyright 2013 American Chemical Society

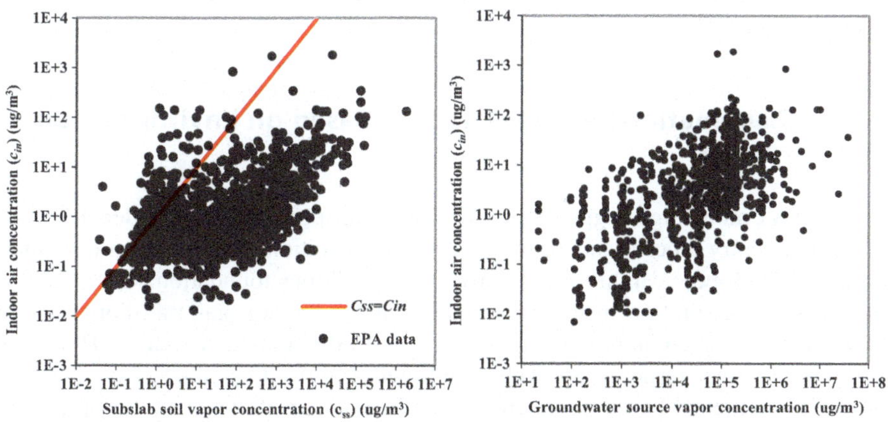

(a) (b)

Fig. 3.3 Measured indoor air concentration data from the US EPA's VI database [6]. (**a**) measured indoor air concentration as a function of subslab concentration; (**b**) measured indoor air concentration as a function of calculated groundwater source vapor concentration. Reprinted with permission from [4]. Copyright 2013 American Chemical Society

threshold, about several hundred μg/m^3, the c_{in} is observed to be linear with c_{ss}, consistent with the modeling expectation. In Fig. 3.3b, the measured c_{in} again is shown as a function of calculated c_{s}. It should be noted that though based on the same database, as c_{s} and c_{ss} were not measured together every time in the database, a full comparison of all site data cannot be offered. Figure 3.3b shows that there is only a very weak dependence of c_{in} on c_s. Compared to the great variation in c_s and c_{ss} (seven orders of magnitude), most of c_{in} are within a much narrower range of two orders of magnitude. Thus, the suspicion was confirmed that there must be reasons to keep c_{in} from changing linearly with c_s or c_{ss}, as excepted by standard VI modeling [4].

The lower limit in Fig. 3.3a might be explained by the detection limit (DL), which is typically between 0.1 and 1 μg/m^3 for chlorinated chemicals. It is reasonable as any indoor measurements lower than this would not be reported as evidence of VI. On the other hand, the upper limit is about 10 μg/m^3, which is still too low for residents to respond. A possible reason behind this might be the presence of indoor adsorption equilibria [4]. Another possible explanation could be that the measured indoor air concentrations are the results of background sources instead of the subsurface ones. Most background levels of chlorinated chemicals are lower than such limits [5]. The influences of background sources will be discussed later in this chapter.

The scatter of two orders of magnitude in measured c_{in} could be caused by the temporal and spatial uncertainties of the indoor environment. For uncontrolled indoor environment, such uncertainties make it unrealistic to obtain precise indoor pressure conditions and indoor air exchange rates, and thus it is impossible to predict the accurate attenuation factors of c_{in} relative to any subsurface contaminant soil gas concentrations [4].

3.3 Examinations of the Database Based on Environmental Factors

Standard VI modelings suggest that soil vapor transport should be independent of the soil gas entry process and building operation conditions [7–9], except in a small porous or high-permeability region around the building foundation. Compared to indoor air concentration inside occupied buildings, the soil gas transport is more steady and affected mainly by natural factors instead of human activities. Thus, it is possible and much easier to use well-established methods to simulate the influences of those key factors based on determined governing equations [10]. Theoretically, site investigation data involving soil, soil gas, and groundwater can and should be consistent with measurements of agreed-upon specific environmental factors, except for biodegradation, which is too unpredictable [4].

With the help of the ASU model (see Chap. 2) [7, 11, 12], US EPA in 2012 released a technical report entitled, "Conceptual model scenarios for the vapor intrusion pathway" [13], in which a summary of environmental factors was presented, as shown in Table 3.1. Those factors can be divided into the vapor source, soil, and building conditions. Leaving out indoor depressurization and air exchange rate, the rest are included in the modeling to simulate soil vapor transport in typical scenarios. Many factors can potentially influence soil vapor transport in VI, and it is generally difficult or impossible to include most of them in a specific field investigation. So it is crucial and also efficient to identify the most critical factors before

Table 3.1 The summary of influences of environmental factors in normalized subsurface contaminant soil concentration profile (oom refers order(s) of magnitude) [13]

	Environmental factors	Influence
Building	Building foundations (basement, crawl space and slab-on-grade);	Contributes <1 oom for scenarios with uniform source and soil, supported by [5, 12, 15] $\frac{c_{ss}}{c_{gw}} \approx \sqrt{\frac{d_f}{d_s}}$ [15]
	Multiple buildings;	Contributes <1 oom, supported by [5, 16]
	Indoor pressurization;	Contributes <1 oom, supported by [5, 12, 15, 17]
	Indoor air exchange rate;	Does not directly affect soil vapor transport
	Impermeable surface cover;	Contributes <1 oom for scenarios with uniform source and soil
	Indoor pressure distribution (wind load);	Contributes <1 oom, as indoor pressurization [5]
Soil	Soil permeability;	Like indoor pressurization, it only changes the soil gas flow rate, which plays an insignificant role in soil gas concentration profile [5, 12, 15]
	Effective diffusivity;	Does not affect soil gas concentration profile for uniform soil
	Layered soil of different effective diffusivity or permeability (horizontal and vertical);	Can play a significant role [5, 18] and depends on the distribution, i.e., the existence of capillary fringe may induce orders of magnitudes effect [19]
	Biodegradation rate (for petroleum hydrocarbon);	Can significantly affect soil gas concentration profile, depending on source concentration and biodegradation rate [5, 7, 11]
	Sorption;	Can induce significant change associated with transients [5]
Contaminant source	Source concentration;	When examines $\frac{c_{ss}}{c_{gw}}$, matters only for scenarios with biodegradation [5, 7, 11]
	Source depth;	Contributes <1 oom for scenarios with uniform source and soil, supported by [5, 12, 15]
	Source-building separation;	For vertical separation, the influence is limited in a uniform soil; for lateral separation, the contaminant concentration decays exponentially with the horizontal transport distance [5, 12, 20, 21]. $\frac{c_{ss}}{c_{gw}} \approx \sqrt{\frac{d_f}{d_s}} e^{-\frac{d_h}{2d_s}\pi}$ [20]
	Multiple sources;	Complex, depends on source concentration and distance

Note: the influences were only considered for non-biodegradation cases unless included in the explanation.

spending considerable effects on them, with the help of modeling [14]. Table 3.1 summarizes the quantified influences of those environmental factors on soil gas concentration attenuations by previous modeling work. It should be noted that Table 3.1 only includes factors in the classical VI scenario and does not include the preferential pathway, which will be discussed in Chap. 5.

3.4 The Generic Attenuation Factors and Influences of Background Sources

The above discussions in this chapter are based on a classical VI scenario and assume that the measurements in EPA's VI database can represent the real threat of VI for buildings of concern. However, recent studies indicate that the measured indoor air concentration may be significantly influenced by the background sources, such as paint and glues. In the EPA's VI database report released in 2012, source strength screens were applied to obtain the generic attenuation factors by minimizing the influences of background sources [1].

Generic attenuation factors values represent thresholds that are higher than the measured contaminant concentration attenuations factors relative to subsurface soil gas concentrations (such as subslab soil gas and the calculated vapor source of groundwater sources) to indoor air in most cases (i.e., 95%). In such a way, applying generic attenuation factors can provide a conservative estimate of indoor air concentration more easily with almost no site characterization work, which is necessary to employ screening models.

In acquiring the generic attenuation factors, different sets of criteria were applied by the EPA to minimize the influences of indoor or outdoor contaminant sources on indoor air concentration. For example, the paired measurements with contaminant groundwater vapor concentration less than 1000 multipliers of the 90th percentile of background levels and subslab soil vapor concentration less than 50 multipliers were screened out for the generation of generic attenuation factor. As a result, US EPA recommends $c_{in}/c_s = 0.001$ and $c_{in}/c_{ss} = 0.03$ as the corresponding generic attenuation factors, as they are the 95th percentile level of the remaining datasets [1].

Yao et al. [22] in 2015 provided an alternative to generating generic attenuation factors of subslab-to-indoor air by making a fitting curve of the plot showing the dependence of indoor air concentration on subslab concentration. Figure 3.4 shows the influence of observed contaminant subslab concentration on observed contaminant indoor air concentration in US EPA's VI database. In this figure, the blue line represents where the contaminant subslab and indoor air concentrations are the same, which should not be expected in a theoretical VI scenario. It seems that most c_{ss} fall to the right of the blue line, indicating c_{in}/c_{ss} is lower than one due to dilution. In a few cases, c_{ss} is lower than c_{in}, possibly caused by the presence of other sources (outdoor air, sink, building materials, etc.) or inappropriate source charactereization (which will be discussed later in this chapter). As discussed above, the lower limit (horizontal dashed blue line) for the observed indoor air concentrations of ±0.1 μg/m^3 could be a detection limit [4].

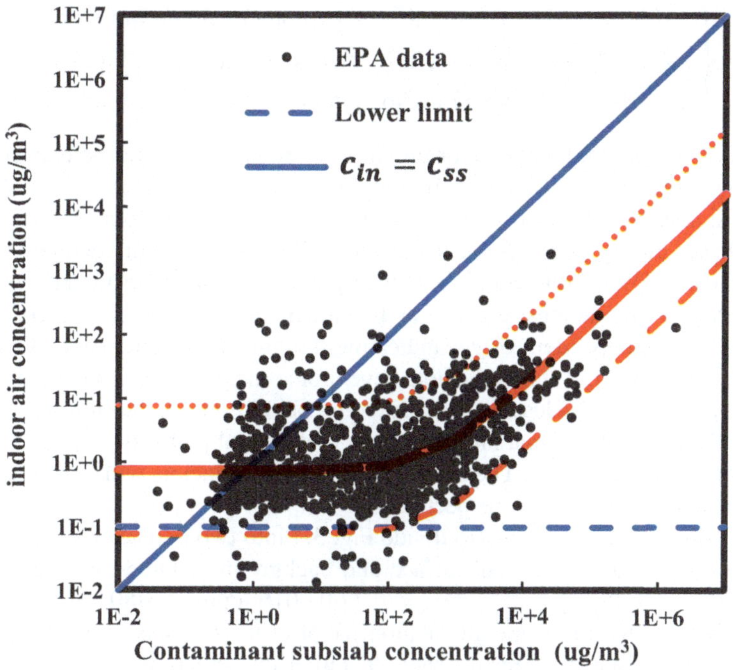

Fig. 3.4 The influences of measured contaminant subslab concentration on measured indoor air concentration in US EPA's VI Database. (The fitting curve was made based on log(c_{in}) instead of c_{in}). Reprinted with permission from [22]. Copyright 2015 Elsevier B.V

In Fig. 3.4, when the contaminant subslab concentration is less than a threshold level which is about 500 μg/m^3, c_{in} is within a range from 0.1 to 10 μg/m^3, indicating the potential role of background [23]. When the c_{ss} is higher than that threshould, c_{in} is linear to c_{ss}. This might suggest a dominnant influence of subsurface sources, as pointed out in the second section of this chapter. Regardless of the possible reasons behind this weird phenomenon, a fitting curve can be made by employing log(c_{in}) instead of c_{in} as the dependent variable, and the equation of the curve can be expressed in the following form [22].

$$c_{in} \approx 1.6 \times 10^{-3} c_{ss} + 0.77\ \mu g/m^3 \tag{3.1}$$

where 0.77 μg/m^3 could be considered as the logarithmic average of the background regardless of contaminant types, while 1.6×10^{-3} is the logarithmic average of the subslab-to-indoor air concentration attenuation factor. The R^2 of this equation for log (c_{in}) is 0.33.

If Eq. (3.1) is multiplied by 10 and 0.1, the maximum and minimum estimates of c_{in} can be calculated based on c_{ss}, respectively [22].

$$c_{\text{in}} = \begin{cases} 1.6 \times 10^{-2} c_{\text{ss}} + 7.7\ \mu\text{g/m}^3, & \text{max (dotted red line in Fig.3.4)} \\ 1.6 \times 10^{-3} c_{\text{ss}} + 0.77\ \mu\text{g/m}^3, & \text{med(solid red line in Fig.3.4)} \\ 1.6 \times 10^{-4} c_{\text{ss}} + 0.077\ \mu\text{g/m}^3, & \text{min (dashed red line in Fig.3.4)} \end{cases} \quad (3.2)$$

With a positive indoor-outdoor pressure difference, c_{in} could be expected to be even lower than that predicted by the minimum case in Eq. (3.2).

The influences of subsurface and surface or background sources can be determined based on Eq. (3.1). As discussed above, Fig. 3.4 can be divided into two parts based on the subslab concentration. In the right part, c_{in} increases with c_{ss}, consistent with normal VI modelings, suggesting the influences of subsurface sources, while in the left, c_{in} is independent of c_{ss}, indicating the potentially dominant role of background sources [22]. It should be noted even for some paired measurements with the level of measured c_{in} is less than measured c_{ss} but close to it, the subsurface sources may still play a role, potentially through a preferential pathway or due to inappropriate site characterization. The influences of these two factors will be discussed later in this chapter and Chap. 5, respectively.

The critical threshold of c_{ss} to divide Fig. 3.4 into two parts can be calculated by assuming the influences of subsurface and background sources are the same, i.e., $1.6 \times 10^{-3} c_{\text{ss}} = 0.77\ \mu\text{g/m}^3$ and $c_{\text{ss}} = 0.77/0.0016\ \mu\text{g/m}^3 \approx 480\ \mu\text{g/m}^3$. Figure 3.4 shows the distribution of the attenuation factor $c_{\text{in}}/c_{\text{ss}}$ datasets with c_{ss} higher than $480\ \mu\text{g/m}^3$. The statistical results show that after the screen, for less than 5% of the houses, $c_{\text{in}}/c_{\text{ss}}$ are higher than 0.017, similar to the recommended 0.03 also based by US EPA based on other source strength screens [22] (Fig. 3.5).

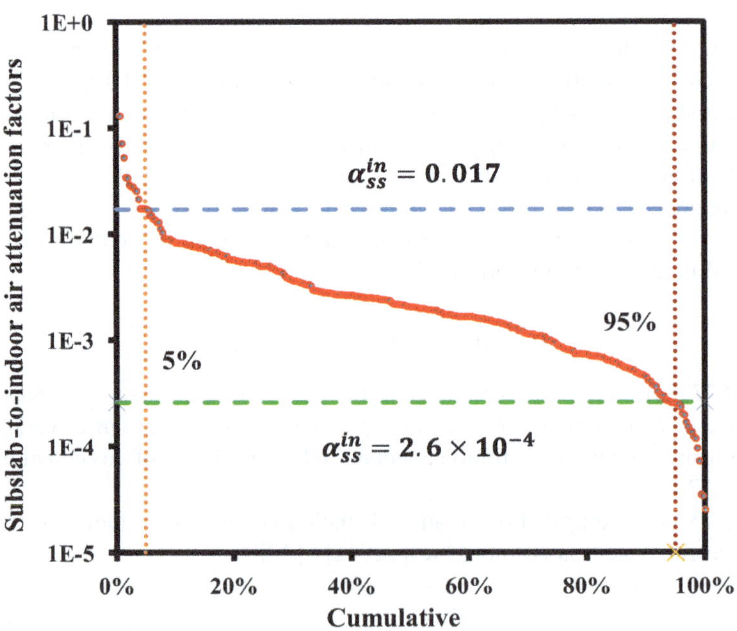

Fig. 3.5 Distribution of subslab-to-indoor air attenuation factors for all residences in US EPA's VI database with subslab concentration above $480\ \mu\text{g/m}^3$. Reprinted with permission from [22]. Copyright 2015 Elsevier B.V

3.5 The Importance of Appropriate Source Characterization

Here the influences of source characterization in the database are discussed. Due to the difficulty of taking groundwater samples directly beneath the building, monitoring wells are usually located some distance away from the target building. In the USEPA's VI database [6], also in most site investigations, the same data from single monitoring well were often used to characterize the VI source for multiple nearby buildings. This practice is because the vapor source is assumed to be very large and with a uniform concentration in a lateral direction. However, it is rare for such an assumption to be met in reality.

For example, in one instance of the datasets in the database, 73 measurements of indoor air contaminant concentrations were linked to a single groundwater measurement. In another instance, 15 indoor measurements were traced back to single monitoring well. On the other hand, 347 groundwater values (161 measured and 186 interpolated) were linked to only one indoor measurement. In sum, there are 427 paired sets of measurements where the same groundwater data was used for multiple buildings, even after the source strength screen. The figure of 427 accounts for more than half of the paired measurements (774) used by the US EPA to generate the generic attenuation factor of groundwater-to-indoor air [1, 24]. Without enough evidence to justify the homogeneity in the groundwater source plume, it is reasonable to question the reliability of this groundwater-to-indoor air attenuation factor (i.e., 0.001) based on such poor source characterization. Even for the 161 cases with one-to-one groundwater-to-indoor air paired measurements, there have been many reported cases [25, 26], demonstrating that it is dangerous to employ measurements from single monitoring well to characterize vapor source, and the records in the US EPA's VI database could be no exceptions for this. Unless the uncertainties in source characterization were minimized, the generic attenuation factor of groundwater-to-indoor air proposed by US EPA remains in question [24].

Figure 3.6 further summarizes horizontal separation distances between groundwater wells and the corresponding buildings. It covers over 500 paired measurements among 774 datasets used to generate the groundwater-to-indoor air attenuation factor. Results show that in less than 5% of the cases, the groundwater samples were taken beneath the building, while in more than 70% of the cases, the samples were taken from monitoring wells located more than 30 m away from the target buildings. In one case, the employed monitoring well was reported to locate over 200 m away from the building of concern. It is reasonable to believe that the site investigators often used the groundwater sample from the nearest monitoring well to produce the source data [24]. Therefore, the groundwater concentrations recorded in the database were actually obtained regardless of the separation distance between the monitoring well and the building with identified VI threat [24]. This however is against the principle recommended by the US EPA itself about a maximum relevant source-building separation distance of 30 m in the lateral direction [10, 27]. So, a substantial fraction of the datasets should not be considered reliable for establishing the subsurface source strength and included in generating the generic attenuation factor based only on a presumption of a uniform source. Similarly, regulators

$$\frac{c_{\text{in}}}{c_{\text{s}}} = \frac{D_{\text{e,t}} A_{\text{B}}}{L_{\text{T}} V_{\text{b}} A_{\text{e}}} \tag{3.4}$$

where L_T is the vertical separation distance between the building foundation and groundwater source, V_b is the building volume, A_e is the indoor air exchange rate, A_B is the building foundation area. $D_{e,\,t}$ is the total effective diffusivity which could be calculated from the following equation [29]:

$$D_{\text{e,t}} = \frac{L_{\text{T}}}{\int_0^{L_{\text{T}}} \frac{1}{D_{\text{e}}(z)} dz} \tag{3.5}$$

where z is the vertical coordinate ($z = 0$ at the water table), and $D_e(z)$ can be estimated using Millington and Quirk [37] equation:

$$D_{\text{e}}(z) = D_{\text{g}} \frac{(\theta_{\text{t}}(z) - \theta_{\text{w}}(z))^{3.33}}{\theta_{\text{t}}(z)^2} + D_{\text{w}} \frac{\theta_{\text{w}}(z)^{3.33}}{\theta_{\text{t}}(z)^2} \tag{3.6}$$

where D_g and D_w are the diffusivity of the contaminant in the gas and water phase, respectively, $\theta_t(z)$ and $\theta_w(z)$ are the total and water-filled porosity in soil, respectively. Generally speaking, the diffusivity in the gas phase dominates as it is faster than the diffusion in the water phase by orders of magnitude. In a simplified case with only one soil layer, $\theta_t(z) = \theta_t$ is a constant and $\theta_w(z)$ could be calculated using Van Genuchten [34] equation:

$$\frac{\theta_{\text{w}}(z) - \theta_{\text{r}}}{\theta_{\text{t}} - \theta_{\text{r}}} = \left(1 + (\alpha * z)^N\right)^{1/N-1} \tag{3.7}$$

where α is the point of inflection in the water retention curve where $\frac{d\theta_w}{dz}$ is maximal; θ_r is the residual soil water content, dimensionless; N is the van Genuchten curve shape parameter.

Yao et al. [24] in 2018 proposed that Eq. (3.3) can be employed to calculate the generic attenuation factor by using conservative parameters in a typical VI scenario. Assume the contaminant is TCE, the soil type is sand, that there is a very shallow groundwater source such that $L_T = 0.5$ m, a typical 10 m × 10 m house footprint with a 2.44 m height such that $V_b = 244$ m^3, that the air exchange rate is at the very conservative end of the range recommended by the USEPA [10], $A_e = 0.1$ h^{-1}, and that it is allowed for not only the slab but the foundation walls to communicate vapor with soil [29], such that $A_B = 180$ m^2, and apply Eqs. (3.4) through (3.7)

$$\begin{cases} D_{\text{e,t}} \approx 1.8 \times 10^{-9}\ \text{m}^2/\text{s} \\ \frac{c_{\text{in}}}{c_{\text{s}}} \approx 10^{-4} \end{cases} \tag{3.8}$$

Unless parameters obtained from a site investigation are more conservative than the employed values, the factor (10^{-4}) represents the physical upper limit of the concentration attenuation factor of TCE from groundwater attenuation to indoor air in a traditional VI pathway (i.e., without preferential pathway). Considering the diffusivities of chlorinated chemicals are not very different from each other, this factor would work for them, too [22].

This calculated result is one order of magnitude lower than 0.001 recommended by the US EPA for the generic groundwater attenuation factor. It is reasonable to conclude that the recommended 0.001 is impossible to be met in reality if the vapor source is characterized correctly in the absence of preferential pathways. This also stands for other models [38–41] that employ realistic assumptions for vertical soil moisture profiles (and thus the vitally important role of diffusive resistance of the capillary zone) to capture vapor transport through the soil. Such limits suggest that the influences of inadequate source characterization or even preferential pathway were likely present in establishing the database that served as the basis for the recommended value of 0.001 [22].

References

1. Dawson H, Kapuscinski R, Schuver H. EPA's vapor intrusion database: evaluation and characterization of attenuation factors for chlorinated volatile organic compounds and residential buildings, EPA 530-R-10-002. Washington, DC: Office of Solid Waste and Emergency Response (OSWER); 2012.
2. US EPA. Draft US EPA's vapor intrusion database: preliminary evaluation of attenuation factors. Washington, DC: Office of Solid Waste and Emergency Response (OSWER); 2008.
3. US EPA. Vapor Intrusion Resources. 2021. https://www.epa.gov/vaporintrusion/vapor-intrusion-resources.
4. Yao Y, Shen R, Pennell KG, Suuberg EM. Examination of the US EPA's vapor intrusion database based on models. Environ Sci Technol. 2013;47(3):1425–33.
5. ITRC. Use of risk assessment in Management of Contaminated Sites, RISK-2. Washington, DC: Interstate Technology and Regulatory Council, Risk Assessment Resources Team; 2008.
6. US EPA. Vapor Intrusion Database. 2021. https://www.epa.gov/vaporintrusion/vapor-intrusion-database.
7. Abreu LDV, Johnson PC. Effect of vapor source– building separation and building construction on soil vapor intrusion as studied with a three-dimensional numerical model. Environ Sci Technol. 2005;39(12):4550–61.
8. Yao Y, Pennell KG, Suuberg EM. Estimation of contaminant subslab concentration in vapor intrusion. J Hazard Mater. 2012;231:10–7.
9. McHugh TE, Beckley L, Bailey D, Gorder K, Dettenmaier E, Rivera-Duarte I, Brock S, MacGregor IC. Evaluation of vapor intrusion using controlled building pressure. Environ Sci Technol. 2012;46(9):4792–9.
10. US EPA. Draft guidance for evaluating the vapor intrusion to indoor air pathway from groundwater and soils (subsurface vapor intrusion guidance). Washington, DC: Office of Solid Waste and Emergency Response (OSWER). EPA530-D-02-004.; 2002.
11. Abreu LDV, Johnson PC. Simulating the effect of aerobic biodegradation on soil vapor intrusion into buildings: influence of degradation rate, source concentration, and depth. Environ Sci Technol. 2006;40(7):2304–15.

12. Abreu LDV. A transient three dimensional numerical model to simulate vapor intrusion into buildings. Doctor of Philosophy,. Tempe, AZ: Arizona State University; 2005.
13. Abreu LDV. Conceptual model scenarios for the vapor intrusion pathway. Washington, DC: Office of Solid Waste and Emergency Response; 2012.
14. Yao Y, Shen R, Pennell KG, Suuberg EM. Examination of the influence of environmental factors on contaminant vapor concentration attenuation factors using the U.S. EPA's vapor intrusion database. Environ Sci Technol. 2013;47(2):906–13.
15. Yao Y, Shen R, Pennell KG, Suuberg EM. Comparison of the Johnson–Ettinger vapor intrusion screening model predictions with full three-dimensional model results. Environ Sci Technol. 2011;45(6):2227–35.
16. Bozkurt O, Pennell KG, Suuberg EM. Simulation of the vapor intrusion process for nonhomogeneous soils using a three-dimensional numerical model. Ground Water Monit Remediat. 2009;29(1):92–104.
17. Muchitsch N, Van Nooten T, Bastiaens L, Kjeldsen P. Integrated evaluation of the performance of a more than seven year old permeable reactive barrier at a site contaminated with chlorinated aliphatic hydrocarbons (CAHs). J Contam Hydrol. 2011;126(3–4):258–70.
18. Abreu LDV, Ettinger R, McAlary T. Simulated soil vapor intrusion attenuation factors including biodegradation for petroleum hydrocarbons. Ground Water Monit Remediat. 2009;29(1): 105–17.
19. Davis GB, Rayner JL, Trefry MG, Fisher SJ, Patterson BM. Measurement and modeling of temporal variations in hydrocarbon vapor behavior in a layered soil profile. Vadose Zone J. 2005;4(2):225–39.
20. Shen R, Pennell KG, Suuberg EM. Numerical evaluation of the effects of the capillary fringe soil moisture on subslab vapor concentration. Sci Total Environ. 2012;437:110–20.
21. Freitas JG, Barker JF. Monitoring lateral transport of ethanol and dissolved gasoline compounds in the capillary fringe. Ground Water Monit Remediat. 2011;31(3):95–102.
22. Yao Y, Wu Y, Suuberg EM, Provoost J, Ma J, Liu J. Vapor intrusion attenuation factors relative to subslab and source, reconsidered in light of background data. J Hazard Mater. 2015;286:553–61.
23. Pennell KG, Scammell MK, McClean MD, Ames J, Weldon B, Friguglietti L, Suuberg EM, Shen R, Indeglia PA, Heiger-Bernays WJ. Sewer gas: an indoor air source of PCE to consider during vapor intrusion investigations. Ground Water Monit Remediat. 2013;33(3):119–26.
24. Yao Y, Verginelli I, Suuberg EM, Eklund B. Examining the use of USEPA's generic attenuation factor in determining groundwater screening levels for vapor intrusion. Ground Water Monit Remediat. 2018;38(2):79–89.
25. Guo YM, Holton C, Luo H, Dahlen P, Gorder K, Dettenmaier E, Johnson PC. Identification of alternative vapor intrusion pathways using controlled pressure testing, soil gas monitoring, and screening model calculations. Environ Sci Technol. 2015;49(22):13472–82.
26. Luo H, Dahlen P, Johnson PC, Peargin T, Creamer T. Spatial variability of soil-gas concentrations near and beneath a building overlying shallow petroleum hydrocarbon–impacted soils. Ground Water Monit Remediat. 2009;29(1):81–91.
27. US EPA. OSWER technical guide for assessing and mitigating the vapor intrusion pathway from subsurface vapor sources to indoor air. Washington, DC: Office of Solid Waste and Emergency Response (OSWER). 9200.2–154.; 2015.
28. The British Columbia Ministry of Environment & Climate Change Strategy. Protocol 22 for contaminated sites: application of vapour attenuation factors to characterize vapour contamination version 1.0. 2017. https://www2.gov.bc.ca/assets/gov/environment/air-land-water/siteremediation/docs/protocols/protocol_22.pdf.
29. Johnson PC, Ettinger RA. Heuristic model for predicting the intrusion rate of contaminant vapors into buildings. Environ Sci Technol. 1991;25(8):1445–52.
30. US EPA. EPA spreadsheet for modeling subsurface vapor Intrusion. 2021. https://www.epa.gov/vaporintrusion/epa-spreadsheet-modeling-subsurface-vapor-intrusion.

31. US EPA. EPA on-line tools for site assessment calculation. 2021. https://www3.epa.gov/ceampubl/learn2model/part-two/onsite/JnE_lite.html.
32. ASTM E2081–00. Standard guide for risk-based corrective action. West Conshohocken, PA: ASTM International; 2015.
33. Yao Y, Verginelli I, Suuberg EM. A two-dimensional analytical model of vapor intrusion involving vertical heterogeneity. Water Resour Res. 2017;53(5):4499–513.
34. Van Genuchten MT. A closed-form equation for predicting the hydraulic conductivity of unsaturated soils. Soil Sci Soc Am J. 1980;44(5):892–8.
35. Verginelli I, Yao YJ, Suuberg EM. Risk assessment tool for chlorinated vapor intrusion based on a two-dimensional analytical model involving vertical heterogeneity. Environ Eng Sci. 2019;36(8):969–80.
36. Little JC, Daisey JM, Nazaroff WW. Transport of subsurface contaminants into buildings. Environ Sci Technol. 1992;26(11):2058–66.
37. Millington RJ, Quirk JP. Permeability of porous solids. Trans Faraday Soc. 1961;57(8):1200–7.
38. Shen R, Pennell KG, Suuberg EM. Influence of soil moisture on soil gas vapor concentration for vapor intrusion. Environ Eng Sci. 2013;30(10):628–37.
39. Atteia O, Hohener P. Semianalytical model predicting transfer of volatile pollutants from groundwater to the soil surface. Environ Sci Technol. 2010;44(16):6228–32.
40. McCarthy KA, Johnson RL. Transport of volatile organic compounds across the capillary fringe. Water Resour Res. 1993;29(6):1675–83.
41. Yao Y, Wang Y, Zhong Z, Tang M, Suuberg EM. Investigating the role of soil texture in vapor intrusion from groundwater sources. J Environ Qual. 2017;46(4):776.

References

US EPA's PVI Database and Vertical Screening Distances

4

4.1 US EPA's Petroleum Vapor Intrusion Database

There are two major types of volatile organic chemicals involving vapor intrusion (VI), chlorinated solvents, and petroleum products, the VI process of which are usually called chlorinated vapor intrusion (CVI) and petroleum vapor intrusion (PVI), respectively. The major difference between CVI and PVI is that chlorinated solvents are considered non-biodegradable in CVI risk assessments, while the oxygen-limited biodegradation of petroleum hydrocarbons is recommended by US EPA to be included in PVI risk assessments. The EPA's VI database by the Office of Solid Waste and Emergency Response (OSWER) focuses on chlorinated chemicals, and there is also an EPA's PVI database supported by the US EPA's Office of Underground Storage Tanks (UST).

The original database was initialized as an empirical dataset by Davis [1–3] and then extended by US EPA to include measurements from more UST and non-UST sites. With the help of the database, the office of UST can promote the understanding of soil gas concentration attenuations in vertical transport and the release of database report and the technical guide of PVI at UST sites in 2013 and 2015, respectively [4, 5]. Like the US EPA's VI database [6], the PVI database was also implemented in an Excel spreadsheet, which can be downloaded at the EPA website, too. The PVI database collected information from 74 PHC sites across North America and Australia [4] (69, 4, and 1 site in the US, Canada, and Australia, respectively). The recorded information included soil texture (with moisture content and porosity), sampling location and time, ground cover, source type and depth, water table depth, and measurements of soil gas concentration. In total, 4221 paired measurements are included in the PVI database. Among them, 61% involved the source type of light nonaqueous phase liquid (LNAPL or light NAPL) sources, and the vapor source for the rest is in the dissolved phase. Most sites in the database were reported to be involved with gasoline releases, and the measurements include 893 benzene soil vapor records, 655 oxygen soil vapor records, and 829 records with paired benzene soil vapor and groundwater data [4].

Y. Yao, Q. Chen, *Vapor Intrusion Simulations and Risk Assessments*,
https://doi.org/10.1007/978-981-19-2700-3_4

4.2 Vertical Screening Distances

As noted above, petroleum products are usually considered biodegradable in aerobic conditions. However, in the early years, the application of VI screening models, such as the Johnson-Ettinger (J-E) model [7], does not include oxygen-limited biodegradation. As a result, the predictions of the screening models would be pretty conservative, and overly low screening-level concentrations were thus generated [8]. To address this issue, a screening criterion named vertical screening distance was developed based on the vertical transport distance of petroleum vapors in the aerobic zone [4]. It relies on the assumption that substantial attenuation of vapor concentrations can be generated by biodegradation during the vertical soil gas transport in the aerobic zone so that the petroleum soil vapors would not cause unacceptable risks to human health due to PVI. The screening distances were proposed based on the statistical analysis of US EPA's PVI database with the nonparametric Kaplan-Meier method.

US EPA concludes that for dissolved sources, the screening distance for benzene soil vapor concentrations to attenuate lower than 100 $\mu g/m^3$ is 1.6 m, as 97% and 94% of measured subslab benzene concentration for dissolved sources in the database are lower than 100 and 50 $\mu g/m^3$, respectively. For LNAPL sources, the screening distances are 4.1 and 6.1 m for UST and non-UST sites (e.g., fuel terminal, refinery, and petrochemical), respectively [4]. This finding is based on the fact that approximately 95% and 93% of measured benzene concentration for the LNAPL sources in the database are lower than 100 and 50 $\mu g/m^3$, respectively, if at a vertical source-building separation distance of 4.6 m at UST sites, while for non-UST sites, approximately 90% are lower than 50 $\mu g/m^3$ at a distance of 5.5 m [4]. It should be noted that the screening distance refers to the thickness of the soil layer uncontaminated by petroleum LNAPL to allow the aerobic conditions required for the biodegradation of petroleum products. Otherwise, the separation distance method does not work.

4.3 The Capillary Effect in Cases of Dissolve Sources

Compared to the LNAPL sources, a smaller vertical screening distance is required for dissolved sources of petroleum products. There are two reasons to explain this difference. The first obvious explanation is that the strength of a dissolved source is much lower than that of an LNAPL source, and thus it requires less transport distance for soil vapor concentration to attenuate below a safe concentration level. The other factor is the capillary effect accompanied by the dissolved source in groundwater.

The capillary fringe is the soil layer above the groundwater water table filled with high water-filled porosity due to the capillary effect, also increasing the moisture content above the capillary fringe. As a result, the resistance to the upward soil gas diffusion was greatly enhanced so that the soil vapor concentration attenuation of petroleum products was also increased even in the absence of biodegradation. On the other hand, high moisture content profile can also prevent the oxygen from the surface air to migrate into the soil, in theory, decreasing the thickness of the aerobic zone. At last, oxygen-limited biodegradation is commonly considered to occur in the water phase, and the increase in water content can promote microbe activities in aerobic conditions.

Yao et al. [9] employed the Brown model to examine influences of the capillary effect in PVI involving dissolved sources by providing a better understanding of the behavior of hydrocarbon soil gas vapor in the presence of heterogeneous soil moisture and thus the application of vertical screening distance in such scenarios.

4.3.1 Model Validation

The model was first validated by the experimental data by Ma et al. [10], which performed a pilot-scale aquifer experiment to simulate the coupled vertical transport and oxygen-limited biodegradation of methane from the source of dissolved ethanol. In the experiment by Ma et al. [10], the aquifer was 115 cm thick, and the simulated water table was located 45 cm below the surface, which is open to the atmosphere to allow the surface air to migrate into the subsurface. Soil samples collected at different depths were measured for model validations [9].

Figure 4.1 shows the comparison between simulated and observed concentration profiles [10]. The points refer to observed soil gas concentrations of methane and oxygen, while the lines are the predicted vertical profiles with the numerical model. In general, numerical predictions fit the observations quite well. In the soil layer of 0–30 cm deep below the surface, the oxygen concentration decays linearly with the depth, as the concentration of methane is relatively low, limiting the activities of microbes in the unsaturated vadose zone. In the capillary fringe, which locates at about 30–45 cm deep below the soil surface, there are sharp decreases in both methane and oxygen concentrations, indicating the role of methanotrophic activity, which requires the presence of both methane and oxygen. Below the capillary fringe or the water table, there is virtually no vertical variation for methane, as in anaerobic conditions [9].

4.3.2 Examinations for Cases with Source in the Dissolved Phase

The validated model was then used to examine the dissolved source cases recorded in the PVI database by Yao et al. [9]. Figure 4.2 shows the statistical results of the PVI database [4] and vertical soil gas concentration profiles simulated with an open soil surface. The solid lines represent the probabilities for benzene soil vapor concentrations to be less than the threshold (i.e., 100 $\mu g/m^3$) with coarse-grained or fine-grained soil, while the dashed lines refer to simulations employing soil properties of sand and clay. The agreement between the simulations and statistical results suggests the modeling can capture the critical influences of oxygen-limited biodegradation in determining the decay of petroleum hydrocarbon concentrations in

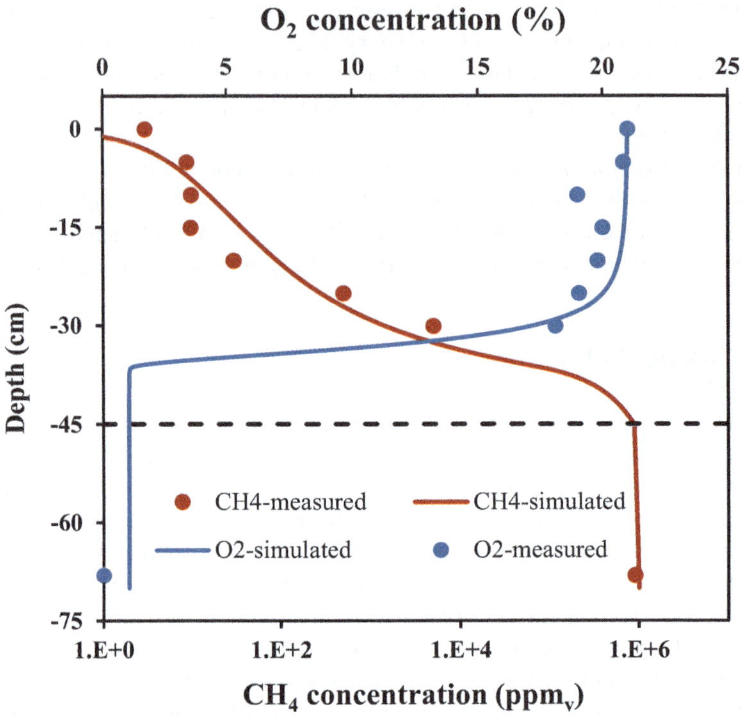

Fig. 4.1 Comparison between predicted concentration profiles (lines) and measured values (points) in lab experiments [10]. Reprinted with permission from [9]. Copyright 2018 Elsevier Ltd

upward transport. Both simulations and the database show that most decay of soil gas concentration occurs at 0–2 and 2–4 ft. above the water table in coarse-grained and fine-grained soil, respectively. Moreover, the total decrease in soil gas concentration is higher in fine-grained soil [9].

The different behavior of the vertical profiles in the near-source zone (i.e., 0–2 ft. above the water table) between two soil types can be explained by the mass balance between the oxygen and hydrocarbon diffusive flux [11].

$$D_{\mathrm{o}}^{\mathrm{eff}}\left[\frac{C_{\mathrm{o}}^{\max}-C_{\mathrm{o}}^{\min}}{\delta}\right]=\mu D_{\mathrm{h}}^{\mathrm{eff}}\left[\frac{C_{\mathrm{h}}^{\max}}{L_{\mathrm{T}}-\delta}\right] \tag{4.1}$$

where $D_{\mathrm{o}}^{\mathrm{eff}}$ is the total effective diffusivity of oxygen across the aerobic zone, $D_{\mathrm{h}}^{\mathrm{eff}}$ is the total effective diffusivity of vapor transport across the anaerobic zone, $C_{\mathrm{o}}^{\max}$ is the maximum oxygen concentration in the aerobic zone, $C_{\mathrm{o}}^{\max}$ is the oxygen concentration in the anaerobic zone, $C_{\mathrm{h}}^{\max}$ is the maximum vapor concentration in the anaerobic zone, δ is the vertical distance from the open surface to the aerobic/anoxic interface, μ is the stoichiometric conversion factor, L_{T} is the distance from the upper surface to the vapor source [9].

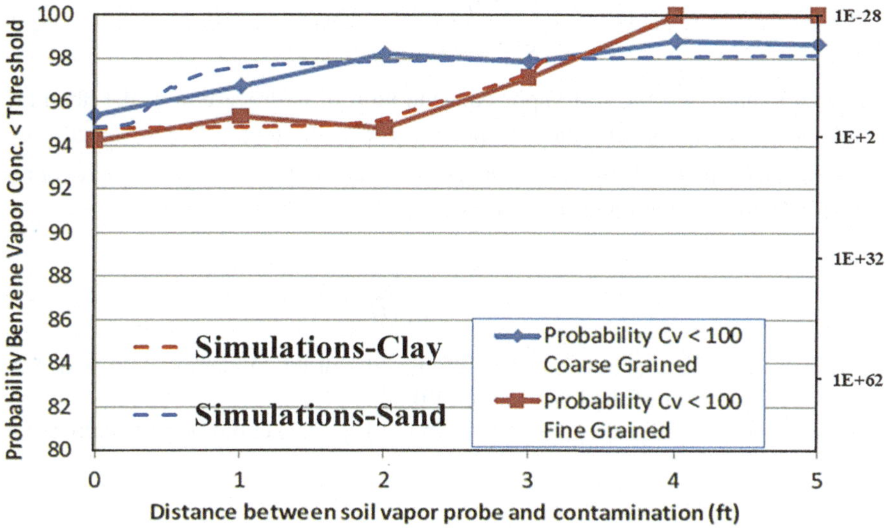

Fig. 4.2 Comparisons of probability (solid lines) for benzene soil vapor concentrations (Cv) to be less than the threshold (100 μg/m^3) for different soil types (coarse- and fine-grained) based on US EPA's PVI database and predicted benzene soil gas concentration profiles (dashed lines) far away from building for the sand and clay cases [4]. In the numerical simulations, the source depth is assumed to be 3 m, the source vapor concentrations are 10 and 5 g/m^3 for the clay and sand cases, respectively, the soil respiration rate is 10^{-7} mol/m^3/s, and the rest of the parameters are shown in the original reference [9]. Reprinted with permission from [9]. Copyright 2018 Elsevier Ltd

Based on the water retention curves calculated with the van Genuchten equation [12], the decline of moisture content with the distance from the water table in coarse-grained soils is usually more significant and fast than that in fine-grained soils. Thus, according to the Millington-Quirk equation [13], the ratio between two effective diffusivities, $D_{\mathrm{o}}^{\mathrm{eff}}/D_{\mathrm{h}}^{\mathrm{eff}}$, in eq. (4.1) should be expected to be higher in coarse-grained soils, as with lower moisture content in the upper unsaturated zone. Comparatively, the effective diffusivity $D_{\mathrm{o}}^{\mathrm{eff}}$ for oxygen in fine-grained soil is lower, making it more difficult to penetrate deep into the soil. As a result, oxygen-limited biodegradation is more easily observed in the near-source zone of coarse-grained soil, causing it more possible for the soil gas concentration to be lower than the threshold [9].

With the increase in the vertical separation distance from the water table, the oxygen also becomes available even in the find-grained soil, allowing the occurrence of oxygen-limited biodegradation [9]. Moreover, as mentioned above, the moisture content in the fine-grained soil is higher than that in the coarse-grained soil, promoting the biodegradation rate, which is usually considered linear to the water content [14].

In a one-dimensional (1D) analytical model developed by DeVaull [15], the vertical soil gas concentration profile of hydrocarbon in the aerobic zone can be estimated using the following equation, based on a piecewise first-order reaction.

$$\frac{c(z)}{c_s} = \frac{(\exp(-\gamma) - \beta) \cdot \exp(\gamma \cdot z/L) + (\beta - \exp(\gamma)) \cdot \exp(-\gamma \cdot z/L)}{\exp(-\gamma) - \exp(\gamma)} \tag{4.2}$$

where $c(z)$ and c_s are the soil vapor concentration of hydrocarbon at the height of z and source vapor concentration, respectively, L is the total transport length, $\beta = \frac{c(z/L=1)}{c_s}$ is the normalized soil vapor concentration at $z = L$, and γ is the square root of diffusive Damkohler number, defined as [9].

$$\gamma = L\sqrt{\frac{r\theta_w}{D_{eff}H}} \tag{4.3}$$

where r is the reaction rate constant of the first-order biodegradation and H is Henry's law constant of the hydrocarbon [9].

By assuming $\beta \approx 0$ and $\exp(-\gamma) \ll \exp(\gamma)$, Eq. (4.3) could be approximated as [9]:

$$\frac{c(z)}{c_s} = \exp\left(-\gamma \cdot \frac{z}{L}\right) = \exp\left(-\sqrt{\frac{r\theta_w}{D_{eff}H}} \cdot z\right) \tag{4.4}$$

According to the Millington-Quirk equation, the increase in moisture content θ_w would cause a decrease in D_{eff} and thus the increase of the square root of diffusive Damkohler number γ [9]. As a result, the decrease of $\exp(-\gamma \cdot z/L)$ is expected, meaning the decrease in soil gas concentration.

4.3.3 Numerical Simulations for Different Soil Textures

To better demonstrate the influences of the capillary effects in different soil textures, Yao et al. [9] performed numerical simulations for two-dimensional (2D) PVI scenarios in two groups. In the first, the vapor source is assumed to be located at the water table, and the source in the other group is located at 1 m above the water table. For each group, three soil types, sand, sandy loam, and clay, represent coarse-grained, medium-grained, and fine-grained soils, respectively.

The simulated soil gas concentration profiles of contaminant and oxygen are shown in Fig. 4.3 for three soil types. In these simulations, the heterogeneous moisture content in the vertical direction was included according to the van Genuchten equation [12]. In all cases, the predicted indoor concentrations in cases including the capillary effect are much lower than those with uniform moisture content by orders of magnitude. It suggests that such vertical moisture content profiles are pretty important in PVI from dissolved sources. Yao et al. [9] concluded that the influences on the coupled soil gas transport and biodegradation could be divided into three parts.

First, the capillary effect would cause the high moisture content above the water table, decreasing the effective diffusivity of hydrocarbons significantly and the

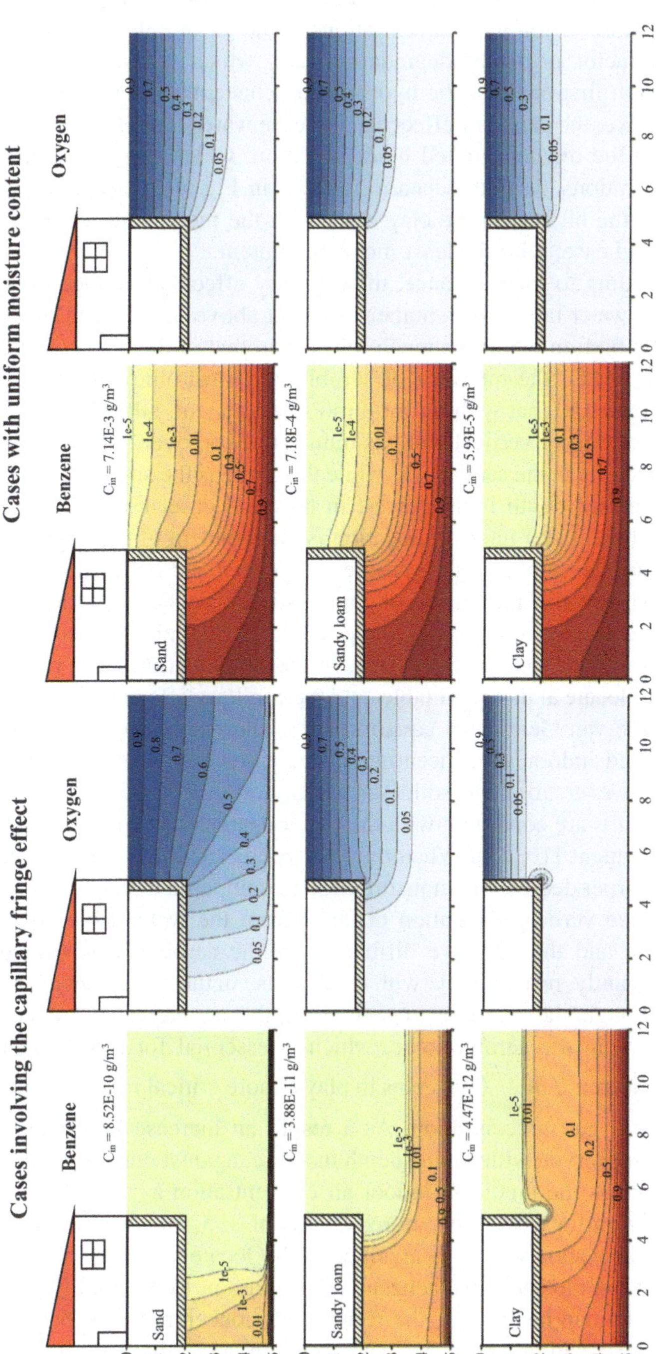

Fig. 4.3 Normalized soil gas concentration profiles of benzene and oxygen for three soil types when source concentration is 100 g/m^3, source depth is 5 m, and vapor source is located at groundwater level. Reprinted with permission from [9]. Copyright 2018 Elsevier Ltd

upward diffusion rate through high-water content and low-permeability layer. As a result, the hydrocarbon soil gas concentrations in the unsaturated zone are lower.

The second factor is the biodegradation rate, which is usually considered a piecewise reaction first-order to the hydrocarbon concentration in the water phase. As discussed above, the capillary effect can increase water content in the unsaturated zone, enhancing the oxygen-limited biodegradation, which can also help decrease soil gas concentrations. In both scenarios shown in Fig. 4.3, the predicted indoor concentration is the highest in the clay case, with the most water content, and the lowest in the sand case, with the least moisture content.

At last, according to mass balance, the capillary effect can induce the moisture content near the water table much higher than that above. As a result, the effective diffusivity at the bottom would be much lower than that in the upper. According to the mass balance [11], the interface of aerobic and anaerobic zones would move downward compared to that with uniform moisture content, allowing more biodegradation activities during vertical soil gas transport. Moreover, the sharpest decline of moisture content is in the sand cases, while the curve is the smoothest for clay. As a result, the interface depth is the lowest in the sand case. For the scenario with uniform water content, the interface location is independent of soil types, as shown by the similar soil gas concentration profiles of oxygen and hydrocarbon.

Figure 4.4 shows the influences of vapor source depth on hydrocarbon and oxygen soil gas concentration profiles and predicted indoor air hydrocarbon concentration for sand and clay cases [9]. Again, in these simulations, the vapor sources were assumed to locate at the water table, and the capillary effects were also included by employing the van Genuchten equation [12]. The results show that in the clay case, the predicted indoor air concentration decreases with the increase of source depth, while, however, for sand soil, it is the opposite. It is not a surprise for the former as the results are consistent with the previous modeling involving a homogenous moisture content [16, 17]. When the soil type is sand, the capillary effect can cause a much sharper decrease in moisture content compared to that in clay soil, with the increase in the vertical separation distance from the water table, and thus the moisture content and the effective diffusivity in the aerobic zone decreases and increases significantly, respectively, with the increase in the source depth. According to Eq. (4.4), though the increase in source depth can induce a longer soil gas transport distance in the aerobic zone, which is essential for biodegradation, the accompanying decrease in $\sqrt{\frac{r\theta_w}{D_{\mathrm{eff}}H}}$ seems to play a more critical role in decreasing the attenuation in soil gas concentration. As a result, an increase in predicted indoor concentration is observed with source depth increase, against our common sense [9].

Figure 4.5 shows the predicted indoor air concentration as a function of source vapor concentration for cases with source depths of 3, 5, and 8 m. In all cases, the upper edge of the vapor source is assumed to be located at the water table. The dashed refers to the threshold for benzene concentration to potentially cause unacceptable risk to human health, i.e., the resident air concentration of benzene in US EPA's Regional Screening Levels [18]. It should be noted that such an assumption does not mean those vapor sources are in the dissolved phase, as in some cases, the

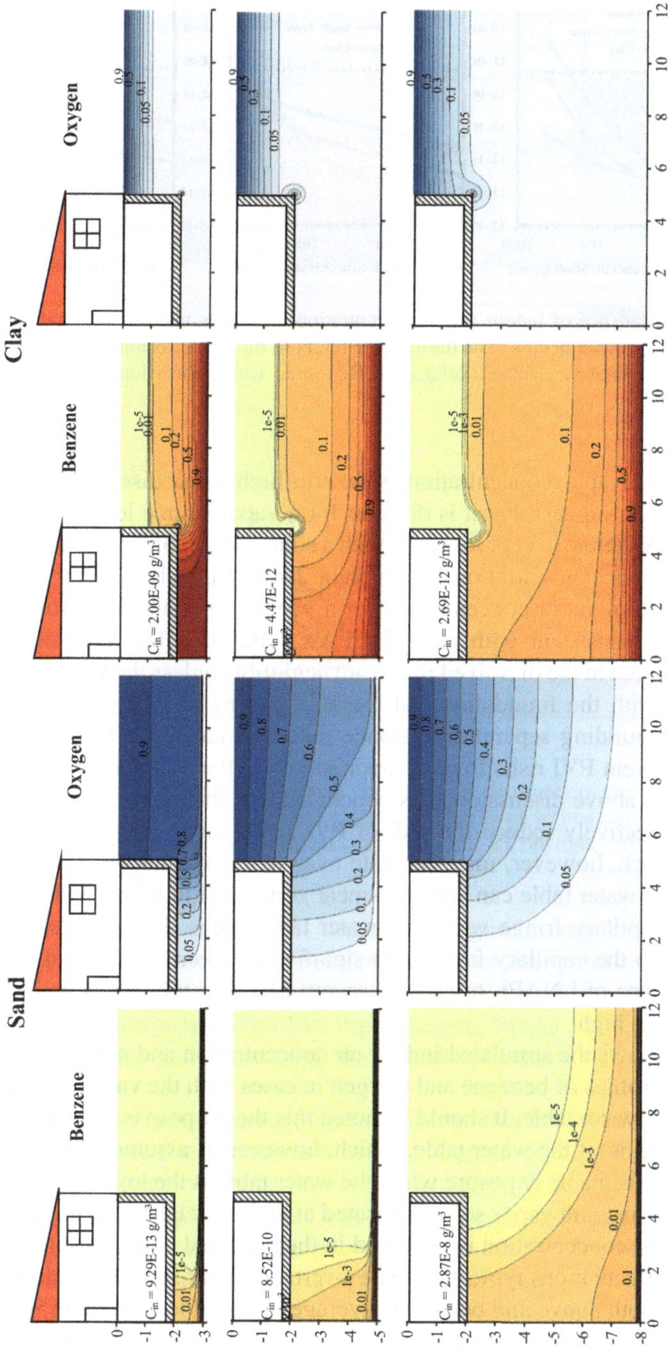

Fig. 4.4 Normalized soil gas concentration profiles of benzene and oxygen for different source depths and soil types when source concentration is 100 g/m^3, and the vapor source is located at the groundwater level. Reprinted with permission from [9]. Copyright 2018 Elsevier Ltd

a vapor source in the dissolved phase, as the smear zone is usually caused by the (residual) LNAPL instead of diluted dissolved source.

4.4 The Role of Soil Texture in NAPL Source Cases

For PVI involving NAPL sources, the capillary effect is no longer critical unless the source is located close to the water table. In cases without the capillary effect, high effective diffusivity in coarse-grained soil is believed to provide less resistance to the upward soil gas flow compared to that of the fine-grained soil. So, in theory, it would increase the PVI risks to human health. However, on the other hand, higher effective diffusivity also means that it is easier for oxygen to migrate into the soil to enhance the oxygen-limited biodegradation, which can cause more attenuation in soil vapor concentration and decrease the risks.

4.4.1 Numerical Simulations

To address this issue, Yao et al. [19] employed the Brown model to explore the influences of soil texture in PVI involving NAPL sources based on the US EPA's PVI database. Figures 4.7 and 4.8 show the simulated soil gas concentration profiles of benzene and oxygen for three soil types (i.e., sand, sandy Loam, and clay) in a slab-on-grade scenario (i.e., the building foundation is 0.2 m deep below the ground surface). In all cases, the vapor source concentration was assumed of 100 g/m^3, and the vapor source is assumed to be located at 3 m (Fig. 4.7) or 8 m (Fig. 4.8) below the ground surface, respectively. Readers interested in the detailed parameters of the simulations are directed to the original reference.

Regardless of soil types, there is an oxygen shadow or an anaerobic zone in the subfoundation in all cases with given parameters, as shown in the right column of Fig. 4.7. Moreover, the sizes of the oxygen shadow or the locations of the aerobic/anaerobic interface are similar in three cases. This is because, in a steady-state, the simulated soil gas concentration profiles are independent of soil types if the moisture content is uniformly distributed, consistent with the results reported by Verginelli et al. [20]. As to the hydrocarbon, the simulated soil gas profiles are the same in the anaerobic zone but quite different in the aerobic zone. This could also be illustrated with Eq. (4.4). The attenuation due to biodegradation along the diffusion distance is determined by the moisture content and also the effective diffusivity. Higher moisture content and lower diffusivity are expected to have a more significant attenuation. Just as the simulations indicate, the highest and the lowest indoor air concentrations are observed in sand and clay cases, respectively, as the former has the lowest moisture content, while the latter has the highest, with a difference of five orders of magnitude. Besides more biodegradation, the lower volumetric soil gas entry rate in the clay case also contributed to such a difference [19].

The soil gas concentration profiles of the 8 m cases in Fig. 4.8. are also similar to those in the 3 m cases. The oxygen soil gas concentration profiles and the

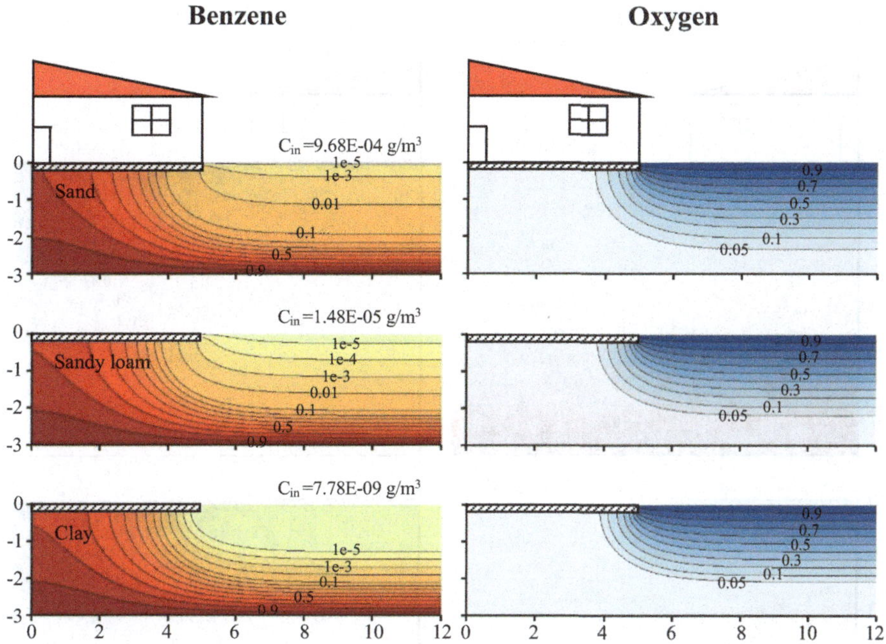

Fig. 4.7 Normalized soil gas concentration profiles for a slab-on-grade scenario with different soil types. A source depth of 3 m below the ground surface and a vapor source of 100 g/m^3 was assumed. Reprinted with permission from [19]. Copyright 2018 The American Society of Agronomy, Crop Science Society of America, and Soil Science Society of America, Inc

hydrocarbon soil gas concentration profiles in the anaerobic zone are independent of soil types, while the highest indoor air concentration was still observed in the sand case, with the lowest in the clay case. Moreover, with the increase in vertical source-building separation distance, the depth of the aerobic/anaerobic interface drops, and the subfoundation zone is under a full aerobic condition [19].

4.4.2 Analytical Analysis

By reorganizing Eq. (4.4), we may recreate a concept named diffusion-reaction length with the following equations [19]:

$$\frac{c(z)}{c_s} = \exp\left(-z/L_R\right) \tag{4.5}$$

where

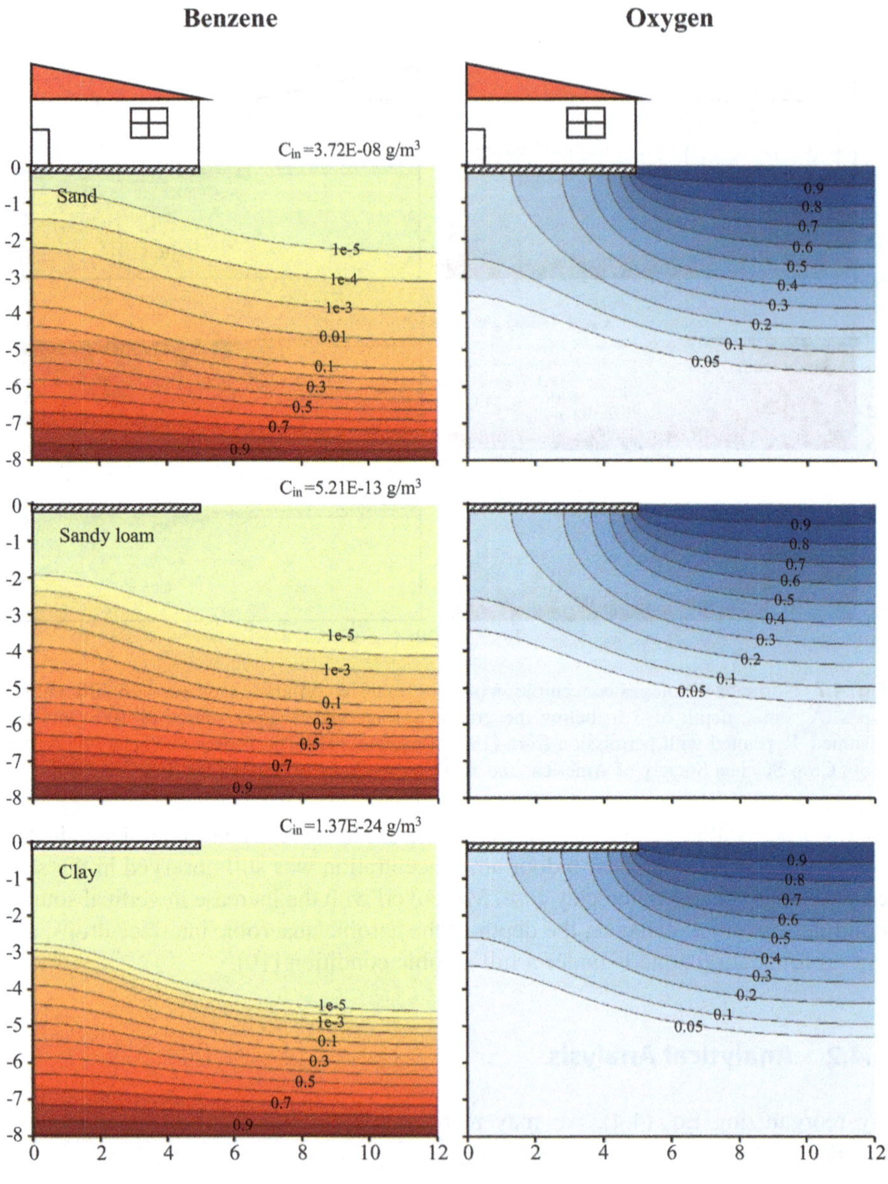

Fig. 4.8 Normalized soil gas concentration profiles for a slab-on-grade scenario with different soil types. A source depth of 8 m below the ground surface and a vapor source of 100 g/m^3 was assumed. Reprinted with permission from [19]. Copyright 2018 The American Society of Agronomy, Crop Science Society of America, and Soil Science Society of America, Inc

$$L_R = \left(\frac{H \cdot D_{\text{eff}}}{r \cdot \theta_w}\right)^{1/2} \tag{4.6}$$

Again, z is the transport distance in the aerobic zone and L_R is the diffusion-reaction length, which can quantify the attenuation of soil gas concentrations due to couped biodegradation and diffusion. The longer this distance is, the less attenuation in soil gas concentration is expected for a specific transport length.

Figure 4.9 shows the diffusion-reaction length (L_R) calculated for the *BTEX* compounds of benzene, toluene, ethylbenzene, and xylenes as a function of the degradation rate constant r (geometric mean $\pm$ standard deviation) of the laboratory and field data discussed by DeVaull [22] and reported by ITRC [21] in cases of three types of soil (Sand, Silt, and Clay). The results show that the diffusion-reaction length is the highest of 15–60 cm in the clay case and the lowest of 3–15 cm in the silt case, positively correlated to the moisture content in the soil. Figure 4.10 further shows the attenuation factor of $\frac{c(z)}{c_s}$ calculated as a function of the diffusion-reaction length (L_R) for different transport distances in the aerobic zone. Again, regardless of the transport distances, the attenuation factor increases with the diffusion-reaction length. It should be noted that a larger attenuation factor refers to less significant attenuation. These analytical calculations further confirm the numerical results about the dependence of soil gas concentration attenuation on soil types for NAPL sources [19].

4.4.3 Refined Vertical Separation Distances Based on Soil Types

Figure 4.11 shows the examination of the US EPA's PVI database [4] with the numerical simulations by the Brown model [19]. Here the solid lines are the probabilities for benzene soil vapor concentrations to be less than the residential screening level of indoor air (i.e., 100 μg/m^3), and the dashed lines refer to numerical simulations of vertical soil gas concentration profiles. In simulations, the soil of sand and clay is used to represent the coarse- and fine-grained soil as recorded in the PVI database. It should be noted that the capillary effect was not included in the simulations, and the soil moisture was assumed to be uniformly distributed. In numerical simulations, the source vapor concentrations in the sand and clay cases were assumed as 50 and 100 g/m^3, respectively, as statistically, the source vapor concentration in coarse-grained cases tends to be lower than that in fine-grained cases in the PVI database.

Both simulations [19] and statistical results [4] show that the soil gas concentrations in the coarse-grained soil tend to decrease at a similar rate in the near-source zone (i.e., within 8 ft. from the vapor source), compared to that in the fine-grained soil. However, above that zone, the decay rate for the fine-grained soil becomes higher than that of the coarse-grained soil. Such a trend suggests that in the near-source zone under anaerobic conditions, the soil gas concentration decays linearly without influences of biodegradation, regardless of soil types, though the

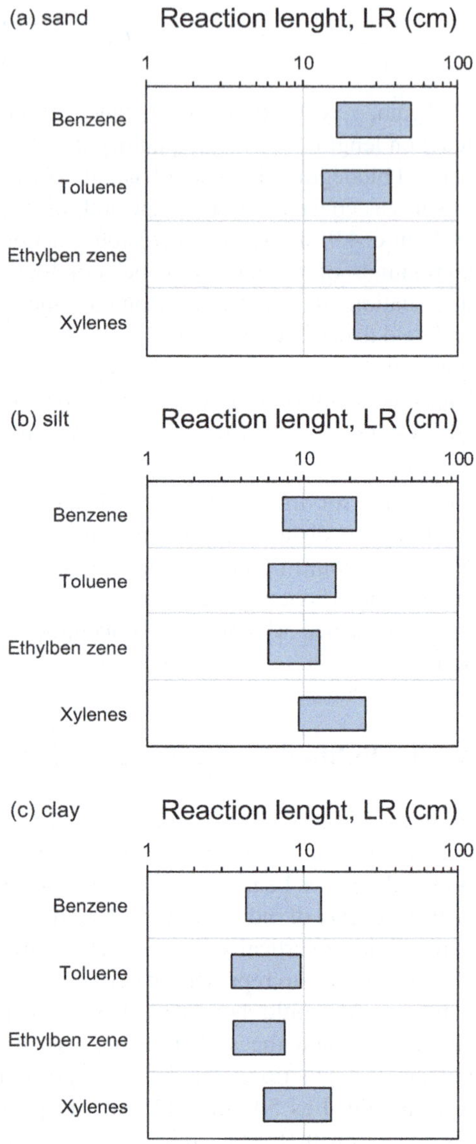

Fig. 4.9 Diffusive reaction length (L_R) calculated for BTEX as a function of the degradation rate constant r (Benzene: $r = 0.087–0.78\ h^{-1}$; Toluene: $r = 0.19–1.4\ h^{-1}$; Ethylbenzene: $r = 0.31–1.4\ h^{-1}$; Xylenes: $r = 0.089–0.64\ h^{-1}$) available in the literature [21] for three types of soil (Sand: $\theta_t = 0.375$, $\theta_w = 0.054$; Silt: $\theta_t = 0.489$, $\theta_w = 0.167$; Clay: $\theta_t = 0.459$, $\theta_w = 0.215$). θ_t: the total porosity. Reprinted with permission from [19]. Copyright 2018 The American Society of Agronomy, Crop Science Society of America, and Soil Science Society of America, Inc

moisture content is higher in fine-grained soils. When biodegradation plays a role in the aerobic layer, the higher moisture content in fine-grained soils would induce smaller diffusion-reaction length and thus a more significant attenuation in soil gas concentration. In other words, with a lower effective diffusivity caused by high moisture content, it takes longer for the hydrocarbon soil gas to transport for a certain distance, allowing more biodegradation to occur, which would bring down the concentration [19].

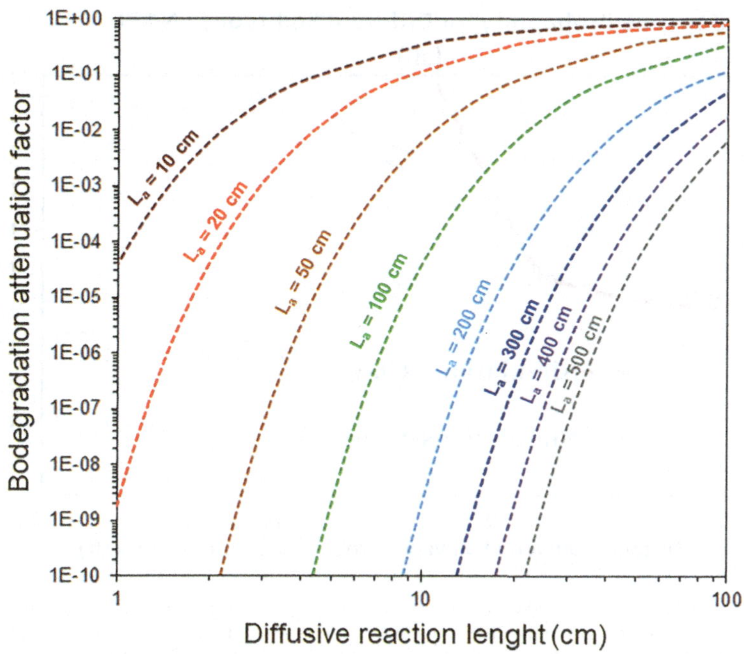

Fig. 4.10 The dependence of attenuation factors on the diffusion-reaction length for different transport distances (La) in the aerobic zone. Reprinted with permission from [19]. Copyright 2018 The American Society of Agronomy, Crop Science Society of America, and Soil Science Society of America, Inc

The role of soil type in determining the soil gas concentration in PVI was further examined with the statistical analysis of US EPA's PVI database [4] in Fig. 4.12, which plots the dependence of vertical exclusion distance on groundwater concentration. The employed datasets are divided into three groups based on soil types, fine-grained, coarse-grained, and very coarse-grained. The results suggest that the vertical exclusion distance can be as large as 5 m in some extreme cases of coarse-grained soils. For fine-grained soils, the maximum exclusion distance is about 3 m, while for the very coarse-grained, the distances are all lower than 1.5 m. It should be noted that compared to datasets of the other two soil types, the number of datasets with very coarse-grained soil is relatively much fewer, and most of them involve a source in the dissolved phase [19].

Based on these results, it is possible to propose refined vertical screening distances according to soil types for LNAPL sources. The new vertical screening distances based on Fig. 4.12 are 5 and 3 m for the coarse-grained and the fine-grained soil, respectively. Due to the insufficient data for the very coarse-grained soil, it might be more appropriate to employ 5 m instead of 1.5 m for the screening distance. These soil type-dependent vertical screening distances are consistent with the analytical modeling shown in Fig. 4.10, which highlights that the lower the diffusive

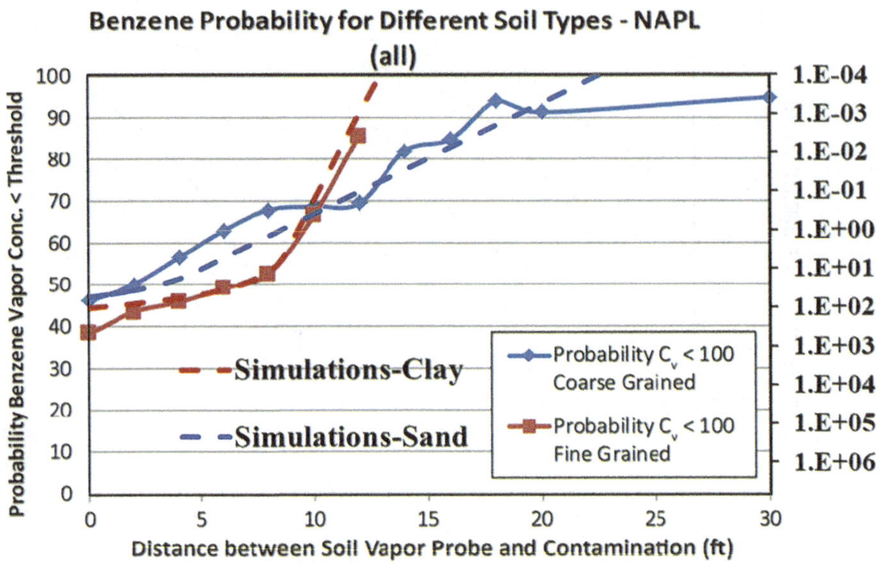

Fig. 4.11 Comparisons of probability (solid lines) for benzene soil vapor concentrations to be less than the threshold (100 μg/m^3) for different soil types (coarse- and fine-grained) based on US EPA's PVI database and predicted benzene soil gas concentration profiles (dashed lines) far away from building for sand and clay cases [4]. Reprinted with permission from [19]. Copyright 2018 The American Society of Agronomy, Crop Science Society of America, and Soil Science Society of America, Inc

reaction length in the fine-grained soil, the more significant attenuation due to biodegradation, and thus the lower the vertical exclusion distance [19].

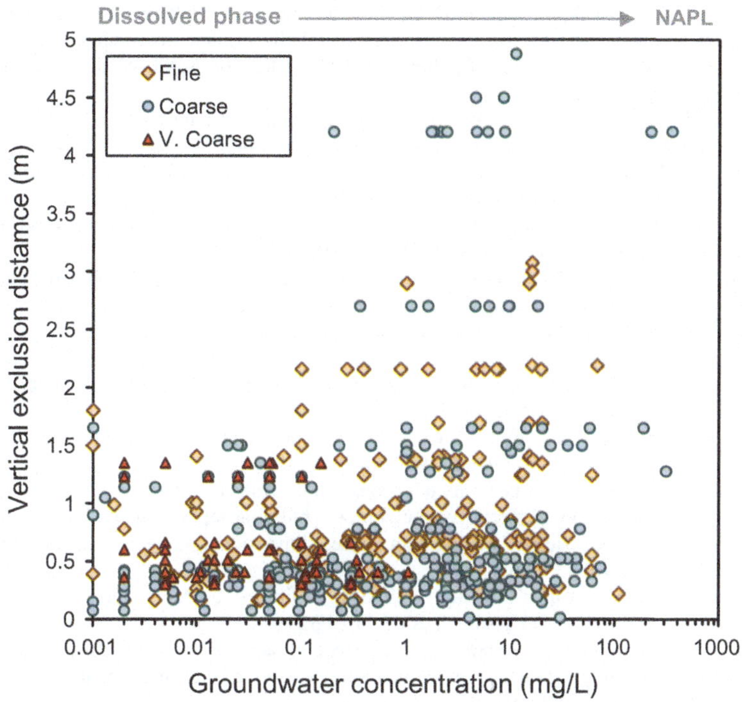

Fig. 4.12 Vertical exclusion distance distances are provided in the US EPA's PVI database [4] for UST and non-UST sites (method 2: Thickness Clean Soil) considering very coarse-grained, coarse-grained, and fine-grained soils. Reprinted with permission from [19]. Copyright 2018 The American Society of Agronomy, Crop Science Society of America, and Soil Science Society of America, Inc

References

1. Davis RV. Bioattenuation of petroleum hydrocarbon vapors in the subsurface: update on recent studies and proposed screening criteria for the vapor-intrusion pathway. LUSTLine. 2009;61: 11–4.
2. Davis RV. Evaluating the petroleum vapor intrusion pathway: studies of natural attenuation of subsurface petroleum hydrocarbons & recommended screening criteria. In: Proceedings of the 21st annual west coast international conference on soil, sediment, water & energy, Mission Valley, San Diego, CA. 2011.
3. Davis RV. Attenuation of subsurface petroleum hydrocarbon vapors: spatial and temporal observations. In: Proceedings of the 21st annual west coast international conference on soil, sediment, water & energy, Mission Valley, San Diego, CA. 2011.
4. US EPA. Evaluation of empirical data to support soil vapor intrusion. Screening criteria for petroleum hydrocarbon compounds, EPA 510-R-13-001. Washington, DC: Office of Underground Storage Tanks; 2013.
5. US EPA. Technical guide for addressing petroleum vapor intrusion at leaking underground storage tank sites. Office of Underground Storage Tanks. EPA 510-R-15-001. 2015.

6. Dawson H, Kapuscinski R, Schuver H. EPA's vapor intrusion database: evaluation and characterization of attenuation factors for chlorinated volatile organic compounds and residential buildings, EPA 530-R-10-002. Washington, DC: Office of Solid Waste and Emergency Response (OSWER); 2012.
7. Johnson PC, Ettinger RA. Heuristic model for predicting the intrusion rate of contaminant vapors into buildings. Environ Sci Technol. 1991;25(8):1445–52.
8. Lahvis MA, Hers I, Davis RV, Wright J, DeVaull GE. Vapor intrusion screening at petroleum UST sites. Ground Water Monit Remediat. 2013;33(2):53–67.
9. Yao Y, Mao F, Xiao Y, Luo J. Modeling capillary fringe effect on petroleum vapor intrusion from groundwater contamination. Water Res. 2019;150:111–9.
10. Ma J, Rixey WG, DeVaull GE, Stafford BP, Alvarez PJJ. Methane bioattenuation and implications for explosion risk reduction along the groundwater to soil surface pathway above a plume of dissolved ethanol. Environ Sci Technol. 2012;46(11):6013–9.
11. Roggemans S, Bruce CL, Johnson PC, Johnson RL. Vadose zone natural attenuation of hydrocarbon vapors: an emperical assessment of soil gas vertical profile data. API Soil Groundwater Res Bull. 2001;15:1–12.
12. Van Genuchten MT. A closed-form equation for predicting the hydraulic conductivity of unsaturated soils. Soil Sci Soc Am J. 1980;44(5):892–8.
13. Millington RJ, Quirk JP. Permeability of porous solids. Trans Faraday Soc. 1961;57(8):1200–7.
14. Abreu LDV, Johnson PC. Effect of vapor source– building separation and building construction on soil vapor intrusion as studied with a three-dimensional numerical model. Environ Sci Technol. 2005;39(12):4550–61.
15. DeVaull GE. Indoor vapor intrusion with oxygen-limited biodegradation for a subsurface gasoline source. Environ Sci Technol. 2007;41(9):3241–8.
16. Abreu LDV, Johnson PC. Simulating the effect of aerobic biodegradation on soil vapor intrusion into buildings: influence of degradation rate, source concentration, and depth. Environ Sci Technol. 2006;40(7):2304–15.
17. Abreu LDV, Ettinger R, McAlary T. Simulated soil vapor intrusion attenuation factors including biodegradation for petroleum hydrocarbons. Ground Water Monit Remediat. 2009;29(1): 105–17.
18. US EPA. Regional screening levels (RSLs). 2021. https://www.epa.gov/risk/regional-screening-levels-rsls.
19. Yao Y, Mao F, Xiao Y, Chen H, Verginelli I, Luo J. Investigating the role of soil texture in petroleum vapor intrusion. J Environ Qual. 2018;47(5):1179–85.
20. Verginelli I, Yao Y, Wang Y, Ma J, Suuberg EM. Estimating the oxygenated zone beneath building foundations for petroleum vapor intrusion assessment. J Hazard Mater. 2016;312:84–96.
21. ITRC. Petroleum vapor intrusion-fundamentals of screening, Investigation, and management. Washington, DC: Interstate Technology and Regulatory Council, Vapor Intrusion Team; 2014.
22. DeVaull GE. Biodegradation rates for petroleum hydrocarbons in aerobic soils: a summary of measured data. In: Presentations at the international symposium on bioremediation and sustainable environmental technologies, Reno, NV. 2011.

Preferential Pathways and the Building Pressure Cycling Method

5

5.1 Introduction of Preferential Pathway and Building Pressure Cycling

In a classical vapor intrusion (VI) scenario, the contaminant soil vapors are released from a subsurface source such as contaminated groundwater and then transport in the vadose zone before reaching the subslab and finally entering the enclosed space of the building through foundation cracks. However, in some circumstances, the soil vapor can migrate into subslab or the indoor space without the soil gas concentration attenuation process during the transport in the vadose zone. For example, it was reported by US EPA that in many cases, soil gas of volatile organic compounds (VOCs) could migrate into the sewer system and then enters indoor spaces as sewer gas [1]. As a result, the VI threat to human health can be more severe than typical cases. In other cases, the land drain or the subsurface pipeline can also serve as a preferential or alternative pathway for VI [2, 3].

There are two types of preferential VI pathways. One promotes the soil gas transport to the subslab zone, such as land drain, while the other helps the contaminant soil gas directly migrate into the indoor space, such as sewer pipeline. The difference between these two is that the foundation crack is the primary pathway for soil gas to enter the indoor space in the former case. These two scenarios are named "pipe flow VI" and "sewer VI," respectively [2]. For the former, Riis et al. [4] reported that the sewer line could perform as a preferential pathway even in the presence of the vapor source plume located far away from the location of buildings threatened by VI. Moreover, they found that the advection of groundwater in the sewer pipeline dominates the vapor migration into the building before the contaminant vaporizes into the sewer air and induces a threat to the indoor air of buildings sharing the same sewer system. Pennell et al. [3] also found that the sewer gas could be a potential source of tetrachloroethylene (PCE) in indoor air. They discovered this after identifying significant spatial variation in contaminant indoor concentrations with the help of sewer gas sampling. As a result, they concluded that the presence of sewer gas in the building could significantly increase the contaminant indoor

Y. Yao, Q. Chen, *Vapor Intrusion Simulations and Risk Assessments*,
https://doi.org/10.1007/978-981-19-2700-3_5

concentration by roughly two orders of magnitude than the cases without a preferential pathway. For the scenario of "pipe flow VI," Guo et al. [2] found that a land drain below the building foundation increases the VI threat and causes three orders of magnitude temporal variations in contaminant indoor air concentration during a long-term on-site monitoring experiment.

Building pressure cycling (BPC) or control pressure method (CPM) is a method of manipulating indoor-outdoor pressure difference and air exchange rate in VI investigations. In practice, there are significant uncertainties in contaminant indoor air concentration, temporally and spatially. Nevertheless, in VI risk assessments, it is recommended by US EPA to obtain maximum reasonable exposure, and indoor air sampling is one of the most common measures. To minimize such uncertainties in indoor air sampling, McHugh and Nickels [5] performed the first field application of BPC by controlling the pressure difference between indoor and outdoor air to achieve maximum contaminant indoor exposure. Later, McHugh et al. [6] improved the BPC application to find the origin of the indoor VOCs, while Beckley et al. [7] employed on-site measurements to obtain instant results of a shorter application of BPC. More recently, a long-term BPC application was carried out at Arizona State University (ASU) Research House in the state of Utah to examine the temporal variations of contaminant indoor and subsurface concentrations. Holton et al. [8] found that BPC can be used to minimize the temporal variations of the indoor concentrations of VOCs by three orders of magnitude, and Guo et al. [2] reported that BPC can be used to confirm the presence of a preferential VI pathway, which is actually a land drain in their study. Recently, Guo et al. [9] performed a series of experiments to adjust the BPC operation parameters to optimize the performance of BPC applications.

5.2 Numerical Simulations of VI Involving Preferential Pathway and BPC

As mentioned above, there have been quite a few field studies reporting that the existence of preferential pathways can significantly affect the VI threat by avoiding the usual concentration through soil gas transport. However, the modeling work to study the soil gas concentration profiles is quite limited, especially under BPC or CPM conditions, enhancing the advective conditions in the subfoundation porous zone.

Yao et al. [10] performed the numerical simulations to investigate the soil gas concentration profiles in VI involving a preferential pathway under normal or BPC conditions. Basically, they tried to repeat the conditions in the long-term field experiment reported by Holton et al. [11] and Guo et al. [2] to validate the model. Then they changed the model parameters (i.e., soil permeability) to examine the temporal behaviors of subsurface soil gas concentrations profiles and indoor air concentrations, with or without a preferential pathway.

In the field study conducted by Holton et al. [11] and Guo et al. [2], a 3–4 year-long experimental was carried out to monitor the temporal variations in contaminant indoor air concentration for a residential house located above groundwater contaminated by chlorinated chemicals, with or without the BPC application and the presence of a preferential pathway. In the simulations by Yao et al. [10], the soil type is set as sandy clay in the whole domain except the subfoundation zone. The vertical variation of moisture content due to the capillary effect was also included in the sandy clay layer, causing inhomogeneous soil permeability and effective diffusivity. To save computation time, only a half domain was simulated, and the location of the opening of the land drain was adjusted to be slightly different from the reality.

5.2.1 Model Validation

Figures 5.1, 5.2, and 5.3 show the observed and simulated indoor-outdoor pressure difference and air exchange rate, contaminant indoor air concentrations, and contaminant emission rates into the house, respectively, during the long-term experiment. In Period #1 (Day 100 to Day 740), the indoor pressure conditions were not controlled (i.e., under natural conditions) and the land drain was not identified yet. For both simulated and measured indoor air concentrations and emission rates, significant temporal variations (up to about three orders of magnitude) in different

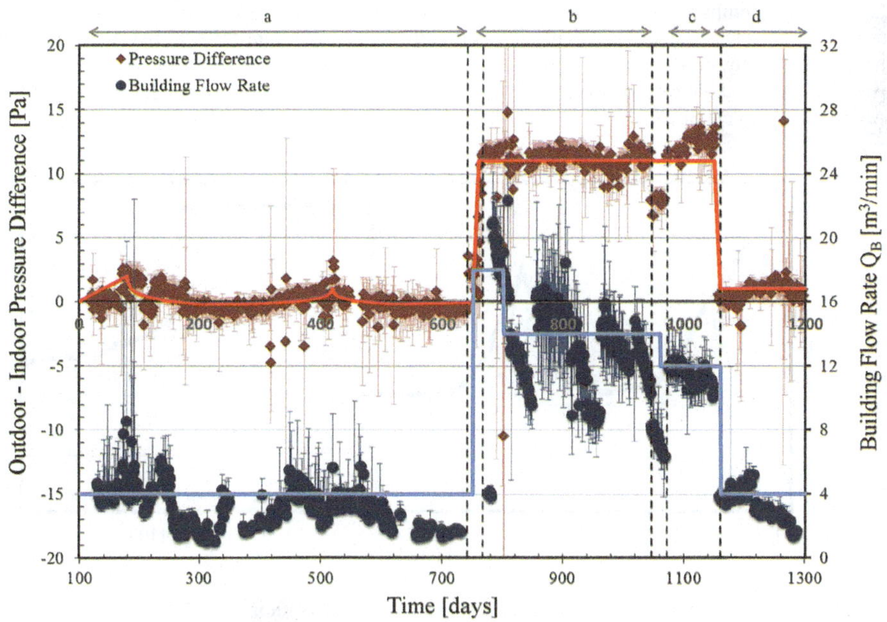

Fig. 5.1 The temporal trends of indoor depressurization and building flow rate (points: measurements; line: simulations). The simulations were reprinted with permission from [10]. Copyright 2017 The American Society of Agronomy, Crop Science Society of America, and Soil Science Society of America, Inc., and the field observations were reprinted with permission from [2]. Copyright 2015 American Chemical Society

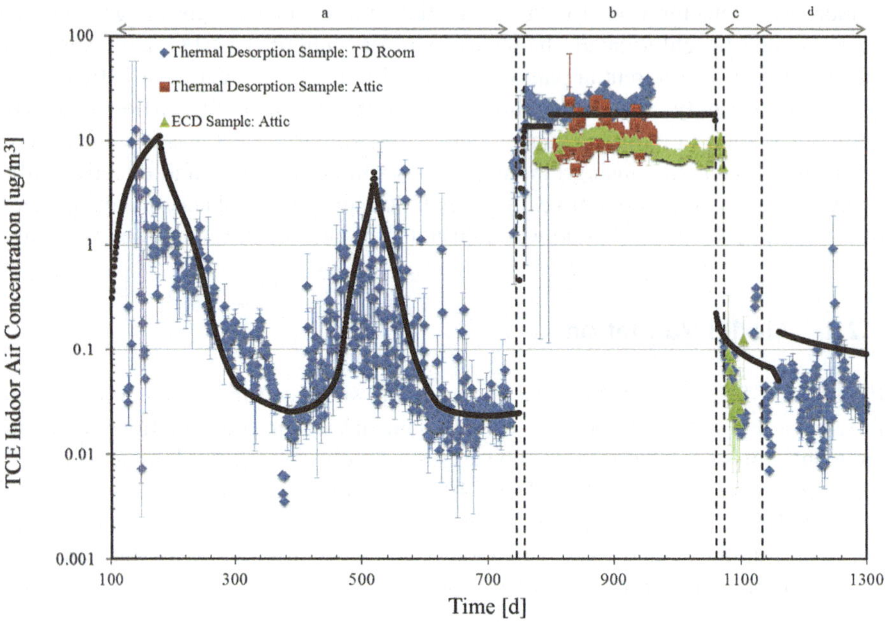

Fig. 5.2 The temporal trends of indoor air concentration (colored: measurements; black: simulation) for TCE. The simulations were reprinted with permission from [10]. Copyright 2017 The American Society of Agronomy, Crop Science Society of America, and Soil Science Society of America, Inc., and the field observations were reprinted with permission from [2]. Copyright 2015 American Chemical Society

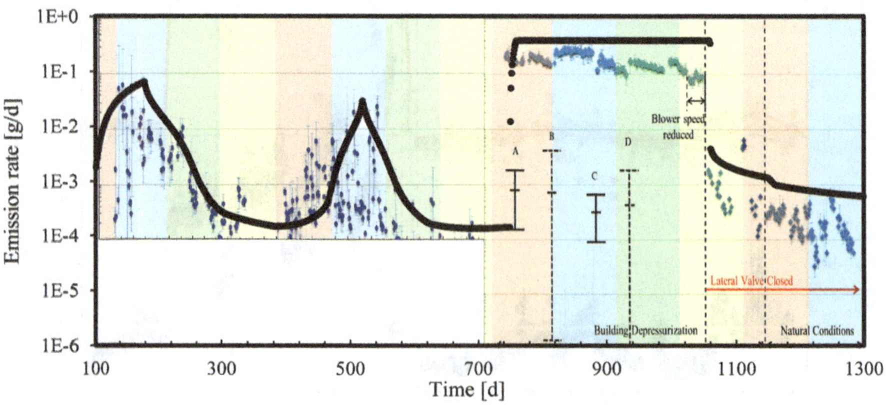

Fig. 5.3 The temporal trends of TCE emission rate into the building (colored: measurements, black: simulation). The simulations were reprinted with permission from [10]. Copyright 2017 The American Society of Agronomy, Crop Science Society of America, and Soil Science Society of America, Inc., and the field observations were reprinted with permission from [2, 8]. Copyright 2015 American Chemical Society

seasons were observed. In contrast, there are no significant variations in the indoor-outdoor pressure difference and air exchange rate. In Period #2 (Day 750 to Day 1050), with an active preferential VI pathway, the indoor pressure was manipulated, and the simulated indoor contaminant concentration and emission rates into the building reached a steady level, then maintained in the rest of this period [10].

In Period #3 (Day 1060 to Day 1170), the land drain was finally identified as a preferential pathway and thus turned off but the indoor pressure was still controlled, the predicted contaminant indoor concentrations and emission rates into the building decreased by nearly two orders of magnitude. After that, the calculated values continued to decline as the subslab contaminant soil gas was depleting, but at a rate relatively lower than observations. A possible explanation might be that the whole subslab zone was assumed as a storage tank in the modeling, while in reality, only a portion played such a role. As a result, the observations were found to decay faster [10].

In Period #4 (Day 1170 to Day 1300), both the land drain and BPC were off, and both simulated and measured variables were still observed to decrease, but contaminant indoor air concentration increased a bit at the beginning due to the immediate drop of the indoor air exchange rate caused by the shut down of BPC. One order of magnitude temporal variation was still observed, while the simulations only showed half. This might be caused by the simplified assumptions of the building flow rate and pressure difference as constants in the simulations, while actually, they were not [10].

5.2.2 The Influences of Soil Permeability and BPC on Indoor Quality

By changing the subfoundation soil permeability, the influences of advection can be examined with a preferential VI pathway. In Figs. 5.4 and 5.5, the simulated contaminant indoor air concentration and emission rate were presented, respectively, with soil permeabilities in the subfoundation as 10^{-8}, 10^{-9}, 10^{-10}, and 10^{-11} m^2. With an active preferential pathway and inactive BPC in Period #1, the simulated fluctuations decrease with the decline in soil permeability, and the maximum levels are approximately linear to the soil permeability. There is a slight fluctuation with the lowest permeability of 10^{-11} m^2, a typical value for the sand soil. However, it should be noted the lower limits of the observations and simulations are the same, regardless of the soil permeabilities [10].

In Period #2, with active BPC and preferential pathway, the indoor air concentration increased linearly to the soil permeability, indicating the role of advection. The simulated indoor air concentration decreases initially before the later rise due to the immediate increase in the air exchange rate [6]. It takes longer to reach a steady-state for cases with lower soil permeability [10].

When the preferential pathway was discovered and thus shut off in Period #3, all the observations and simulations started to drop sharply, even though the indoor-outdoor pressure difference was maintained with BPC. Moreover, in the final stage with inactive preferential pathway and BPC, the simulated and observed indoor air

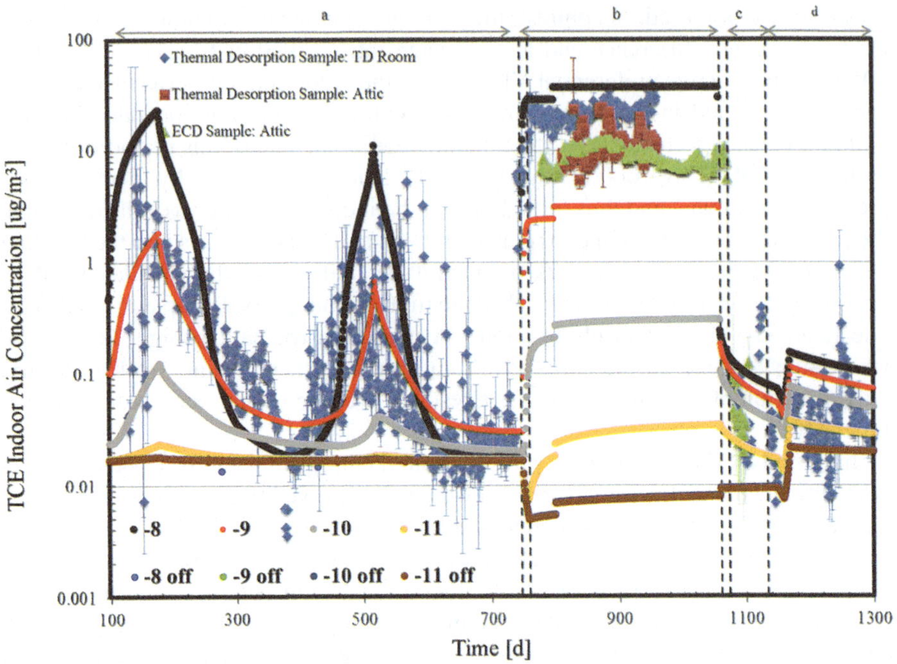

Fig. 5.4 Numerical simulations of indoor air concentration with different soil permeability to soil gas flow in the subfoundation (−8 means the soil permeability to soil gas flow is 10^{-8} m^2 in the subfoundation, and −8 off means the soil permeability to soil gas flow is 10^{-8} m^2 in the subfoundation and the preferential pathway was turned off from the beginning). The simulations were reprinted with permission from [10]. Copyright 2017 The American Society of Agronomy, Crop Science Society of America, and Soil Science Society of America, Inc., and the field observations were reprinted with permission from [2]. Copyright 2015 American Chemical Society

concentrations rose again due to the decreased air exchange rate. It should be noted that the emission rates into the building were still declining, suggesting the slow depletion of the subfoundation contaminant soil gas [10].

The simulations show that the performance of the preferential pathway in "pipe flow VI" depends on the advection and soil permeability in the subfoundation. Its influences can be quite limited if the soil permeability is no higher than typical sand soil. However, it still can play an essential role in the lateral transport of contaminant soil gas without significant attenuation in cases with nonuniform vapor sources, which is a common phenomenon in practice [10].

In comparison, the VI scenarios with an inactive preferential pathway all the time were also investigated. The results suggest that the temporal behaviors are almost identical independent of soil permeability. The application of BPC can even decrease the contaminant indoor air concentrations due to the increase in the air exchange rate, as discussed above. Nevertheless, the emission rate might be slightly increased by an enhanced volumetric flow rate into the building [10].

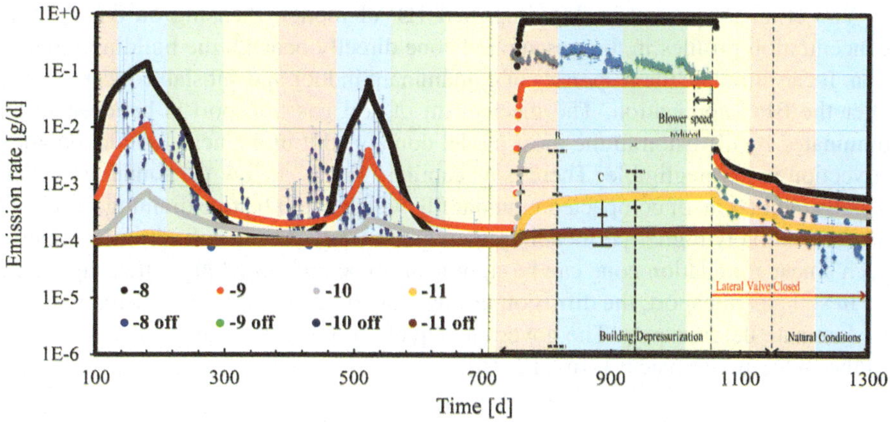

Fig. 5.5 Numerical simulations of trichloroethylene (TCE) emission rate with different soil permeability to soil gas flow in the subfoundation (−8 means the soil permeability to soil gas flow is 10^{-8} m^2 in the subfoundation, and −8 off means the soil permeability to soil gas flow is 10^{-8} m^2 in the subfoundation, and the preferential pathway is off from the beginning). The simulations were reprinted with permission from [10]. Copyright 2017 The American Society of Agronomy, Crop Science Society of America, and Soil Science Society of America, Inc., and the field observations were reprinted with permission from [2, 8]. Copyright 2015 American Chemical Society

This phenomenon can be explained by the classical mass balance in VI developed by Johnson and Ettinger [12]. In their conceptual site model, the contaminant flow rate into the enclosed space is dominated by the upward diffusion in a steady-state. As the BPC application is supposed to affect the advection in a very small zone surrounding the building with high soil permeability, the upward diffusion is unaffected. So the contaminant emission rates into the building would not be increased significantly by indoor depressurization. In fact, Guo et al. [2] also reported little significant difference in contaminant indoor air concentration between Periods #3 and #4. They both are within the range between 0.01 and 0.1 μg/m^3. McHugh et al. [6] also reported similar results that indoor depressurization decreased instead of increasing the contaminant indoor air concentration from subsurface sources at one site.

5.3 The Analytical Solutions for the BPC Performance in the Short Term

In the above section, we discussed the long-term behavior of soil gas concentration profiles and indoor air concentration in the presence of BPC and preferential pathways. However, in practice, investigators usually applied BPC or CPM for a few hours or at most several days to measure a maximum reasonable exposure or identify the existence of a preferential pathway. Thus, they are more interested in the short-term behavior in those scenarios.

Yao et al. [13] developed a simplified 1D VI model, focusing on the soil gas concentration profiles in the unsaturated zone directly beneath the building foundation. It can simulate the responses of contaminant indoor and subslab concentrations after the BPC application. The mechanism of soil gas transport is believed to be dominated by diffusion in the most model domain. But in the near-foundation zone advection is not negligible. There is a "capillary break" below the foundation slab, usually about 4 in. of compacted granular fill, which could be free draining or at least have a relatively higher permeability than natural soil. As a result, the advection in such a near-foundation zone can be significant to be unignored in VI investigations.

In soil gas transport, the diffusion through air-filled porosities is usually assumed to dominate the transport. The governing equation of Ficker's law can be described as the following transient form [12–15]:

$$\frac{\partial^2 c}{\partial z^2} = \frac{a}{D_e}\frac{\partial c}{\partial t} \tag{5.1}$$

$$a = \phi_g + \frac{\phi_w}{H} + \frac{k_{oc} f_{oc} \rho_b}{H} \tag{5.2}$$

$$D_e = D_g \frac{\phi_g^{\frac{10}{3}}}{\phi_T^2} + \frac{D_w}{H}\frac{\phi_w^{\frac{10}{3}}}{\phi_T^2} \tag{5.3}$$

where c is the contaminant vapor concentration [M/L], z is the vertical coordinate [L], a is the retardation factor, defined in Eq. (5.2) ϕ_T is the total soil porosity, ϕ_g is the air-filled porosity, ϕ_w is the moisture-filled porosity, H is the contaminant Henry's Law constant relating vapor phase contaminant concentration to water phase concentration; k_{oc} is the sorption coefficient of contaminant i to organic carbon in the soil [L^3/M]; f_{oc} is the mass fraction of organic carbon in the soil, ρ_b is the soil bulk density [M/L^3], D_g and D_w are the contaminant diffusivities in gas and water phases [L^2/T], respectively, and D_e is the effective diffusivity in porous media [L^2/T], defined in Eq. (5.3) [12]. The transport equation (5.1) is Fick's Second Law with the additional retardation factor. This is a partial differential equation whose consequences are examined with the BPC method [13].

The contaminant indoor concentration c_{in} is time-dependent and could be calculated by considering the indoor space as a well-mixed tank with a volume of V_b being flushed by clean outdoor air (i.e., $c = 0$) at a volumetric flow rate of $AER \cdot V_b$. Then the temporal variation in c_{in} can be estimated by the difference between the contaminant inflow and outflow rate [13], as expressed below [16]:

$$\frac{dc_{in}}{dt} \cdot V_b = -D_e \cdot A \cdot \frac{dc(z = L)}{dt} - c_{in} \cdot \mathrm{AER} \cdot V_b \tag{5.4}$$

where c_{in} is the contaminant indoor air concentration [M/L^3], A is building foundation footprint size [L^2], AER is the indoor air exchange rate [T^{-1}], and V_b is the building volume [L^3]. Figure 5.6 shows two slightly different source configurations. The left part of the figure shows the case with a vapor source (e.g., a NAPL source)

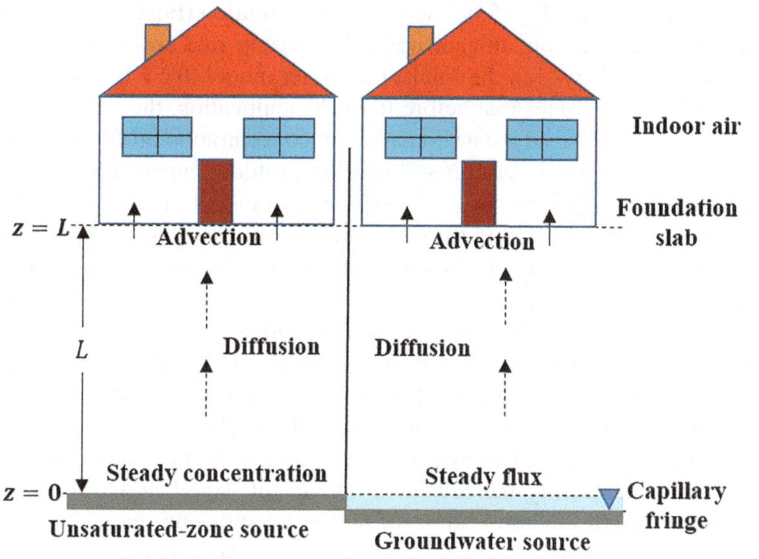

Fig. 5.6 The conceptual model for scenarios of two types of sources. Reprinted with permission from [13]. Copyright 2020 Elsevier B.V

in the vadose zone out of the influence of the capillary effect, while the right part shows a more common groundwater source below the capillary fringe [13].

In the former case, z is the vertical distance from the vapor source, and $z = 0$ is the vapor source with a concentration as c_0. The boundary and initial conditions are expressed as the follows [13]:

$$c(z = 0, t) = c_0 \tag{5.5}$$

$$\frac{\partial c}{\partial z}(z = L, t) = -\frac{Qc(z = L)}{D_e A} \tag{5.6}$$

$$\frac{c(z, t = 0)}{c_0} = 1 - \frac{Bi_1}{1 + Bi_1}\frac{z}{L} \tag{5.7}$$

$$\frac{c(z, t = \infty)}{c_0} = 1 - \frac{Bi_2}{1 + Bi_2}\frac{z}{L} \tag{5.8}$$

$$Bi = \frac{QL}{D_e A} \tag{5.9}$$

where Q is the volumetric flow rate into the building [L^3/T], Bi is the Biot number as defined in Eq. (5.9), using an analogy to the standard heat transfer definition of this quantity [17]. Here Bi refers to the ratio of mass transfer resistance to advection and diffusion at the upper boundary, and subscripts 1 and 2 refer to the cases at $t = 0$ and $t > 0$ (including $t = \infty$), respectively. In this chapter, only indoor depressurization is involved, i.e., $Q_1 < Q_2$. At the upper boundary, i.e., $z = L$, only advection is considered as the only mechanism in determining the contaminant entry rate into the building, as shown in Fig. 5.6. The boundary condition (Eq. 5.6) reflects a forced mass balance between contaminant upward diffusion rate and entry rate into the building, just as employed in the Johnson-Ettinger model (the J-E model) [12]. The initial condition (5.7) states that before the BPC application, the initial condition is applied by assuming a steady-state contaminant concentration profile ($0 < z < L$). In our cases, we employed a uniform soil and the profile is linear. Basically, it is the analytical solution of the classical Johnson-Ettinger model in steady-state [13].

The boundary condition (5.6) is based on the assumption that advection dominates the soil gas entry into the building, which is correct as long as it is indoor depressurization. And it should be noted $Q = Q_2 > Q_1$ at $t > 0$. With the BPC application for an infinite time, the contaminant profiles would achieve a new steady-state, but different from the one before the BPC application. Thus, a new Biot number (i.e., Bi_2) could be employed, as shown in Eqs. (5.7–5.8). So the temporal variations in vertical contaminant profiles after the application is a process moving from one steady-state described by Eq. (5.7) to another by Eq. (5.8). The interesting questions are how quickly the subslab contaminant would move to a new semi-steady-state level and the implications of the observations during the test.

In such a case, the mathematical solution for the transient contaminant concentration profiles can be expressed as follows [13]:

$$\frac{c(z,t)}{c_0} = 1 - \frac{Bi_2}{1+Bi_2}\frac{z}{L} + 2\frac{Bi_2 - Bi_1}{1+Bi_1}\sum \frac{1}{Bi_2^2 + \lambda_n^2 + Bi_2}\exp\left(-\lambda_n^2 \frac{D_e}{a} t\right)\frac{\sin(\lambda_n z)}{\sin(\lambda_n L)} \tag{5.10}$$

$$\tan(\lambda_n L) = -\frac{\lambda_n L}{Bi_2} \tag{5.11}$$

where λ_n, the roots of the transcendental equation (5.11), are the eigenvalues of the equation, dependent on the Biot number [13]. This mathematical solution was not new and could be found in various classic books about heat and mass transfer [17–19].

When the vapor source is contaminated groundwater, it should be noted that a capillary fringe is located above the source, representing a large resistance to upward contaminant diffusion at the water table. This is because the water content in the capillary fringe is very high (i.e., the soil porosity is almost full of water) [13]. Previous studies suggested that the contaminant concentration located above the capillary layer should usually be lower than the calculated soil vapor concentration based on groundwater measurements and Henry's law by one to two orders of magnitude [20, 21].

The above governing equations still work, but the boundary condition at the bottom needs to be changed, as $z = 0$ refers to the location of the top of the capillary fringe instead of the vapor source now, as shown in the right part of Fig. 5.6. Moreover, it is the diffusion through the capillary fringe that determines the vertical transport rate and also the contaminant entry rate in a steady-state (i.e., $t = 0$ or $t = \infty$). In short, the lower boundary condition changes from the first type to the second type, from fixed concentration to fixed flux. The new boundary condition also determines that the contaminant flow rate into the building may fluctuate during the BPC application, but the values are approximately the same or slightly different for the beginning and the end. With indoor depressurization, it cannot be assumed that the contaminant concentration at $z = 0$ is constant as c_0, like what had been assumed earlier in Eq. (5.5). For groundwater source, the new boundary condition at $z = 0$ to replace Eq. (5.5) can be expressed as [13].

$$\frac{\partial c}{\partial z}(z=0,t) = -\frac{Bi_1}{1+Bi_1}\frac{c_0}{L} \tag{5.12}$$

Unlike the first case, this same lower boundary condition is valid all the time as the contaminant flow rate through the capillary zone is assumed unchanged. At the building foundation, like in the first scenario, there is still a forced mass balance between the upward diffusion rate and the entry rate, as governed by Eq. (5.6). The initial conditions can still be described with Eq. (5.7), but a new alternative should replace Eq. (5.8) for the vertical soil gas concentration profile at $t = \infty$ [13].

$$\frac{c(z,t=\infty)}{c_0} = \frac{Bi_1}{Bi_2}\frac{1+Bi_2}{1+Bi_1} - \frac{Bi_1}{1+Bi_1}\frac{z}{L} \tag{5.13}$$

It should be noted that c_0 in Eq. (5.13) is the soil vapor concentration at the top of the capillary fringe instead of the source vapor concentration in the first scenario. The shift in the value of c_0 represents the difference due to the change of source types [13].

The mathematical solution to the problem involving a groundwater source can be expressed as [13].

$$\frac{c(z,t)}{c_0} = \frac{Bi_1}{Bi_2}\frac{1+Bi_2}{1+Bi_1} - \frac{Bi_1}{1+Bi_1}\frac{z}{L} + 2\frac{Bi_2-Bi_1}{1+Bi_1}\sum\frac{1}{Bi_2^2+\lambda_n^2+Bi_2}\exp\left(-\lambda_n^2\frac{D_e}{a}t\right)\frac{\cos(\lambda_n z)}{\cos(\lambda_n L)} \tag{5.14}$$

$$\cot(\lambda_n L) = \frac{\lambda_n L}{Bi} \tag{5.15}$$

where λ_n, the roots of the transcendental Eq. (5.15), are the eigenvalues of the problem, dependent on the Biot number.

5.4 Numerical Simulations for the BPC Performance in the Short Term

The analytically simulated results by Yao et al. [13] suggest that these normalized rates can achieve a pseudo-steady-state value, which is roughly twice the value under natural conditions in the sand case. Nevertheless, 1D models are limited by only considering the vertical soil gas transport and usually assume the diffusion dominant to simplify calculations. They also rely on the mass balance at the upper boundary. As a result, their predictions are often conservative and cannot capture the whole picture of the conceptual site model [22]. To overcome this difficulty, Liu et al. [23] used the Brown model to perform more comprehensive 3D numerical simulations capable of studying the influences of multiple factors in the BPC applications.

5.4.1 Model Validation

Figure 5.7 shows the comparison between the simulations and observations at Building 200 (B200) with different indoor depressurizations [24]. The points and lines refer to measurements and observations, respectively. B200, a medical office of 2100 ft^2 (195 m^2), has a volume of 16,800 ft^3 (476 m^3). It is a single story and has a slab-on-grade foundation. The shallow soil was reported to consist of uniform unconsolidated sediments, most sand and silt. The primary contaminant is TCE [26]. During the BPC application, the building loading rates were measured with indoor pressure manipulated as −10 pa, −20 pa, and −50 pa, respectively [24]. To replicate the actual site conditions, the vapor source is assumed to be very close to the building foundation in simulation [13]. More details about the site investigation and simulations are shown in the original references.

Figure 5.7 shows that the numerical simulations reached the steady-state quickly, and the results were maintained during the investigations. Both simulated and

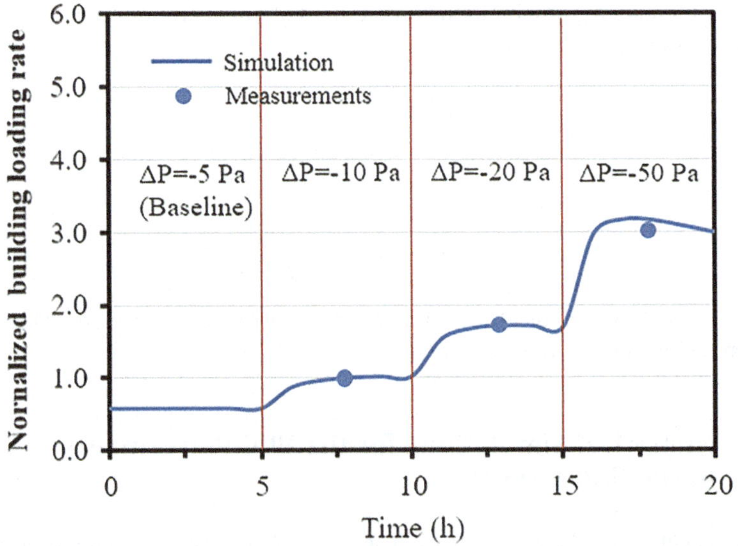

Fig. 5.7 Comparison between predicted normalized building loading rate of 3D numerical models and field experiment at B200 [24]. Reprinted with permission from [25]. Copyright 2020 Springer Nature

measured temporal variations are linear to the pressure difference, indicating the role of advection in determining the loading rate. The difference between the emission rate into the building and building loading rate is that the former refers to the contaminant entry rate into the building while the latter is the escape rate of contaminant from indoor to outdoor. In the steady-state in the long-term evaluation, they are the same, but in the short-term study, the behaviors are a bit different [13].

5.4.2 Influences of Environmental Factors

Figure 5.8 shows the influence of source depth, volumetric entry rates, and organic carbon content of the soil in determining the normalized contaminant entry rate

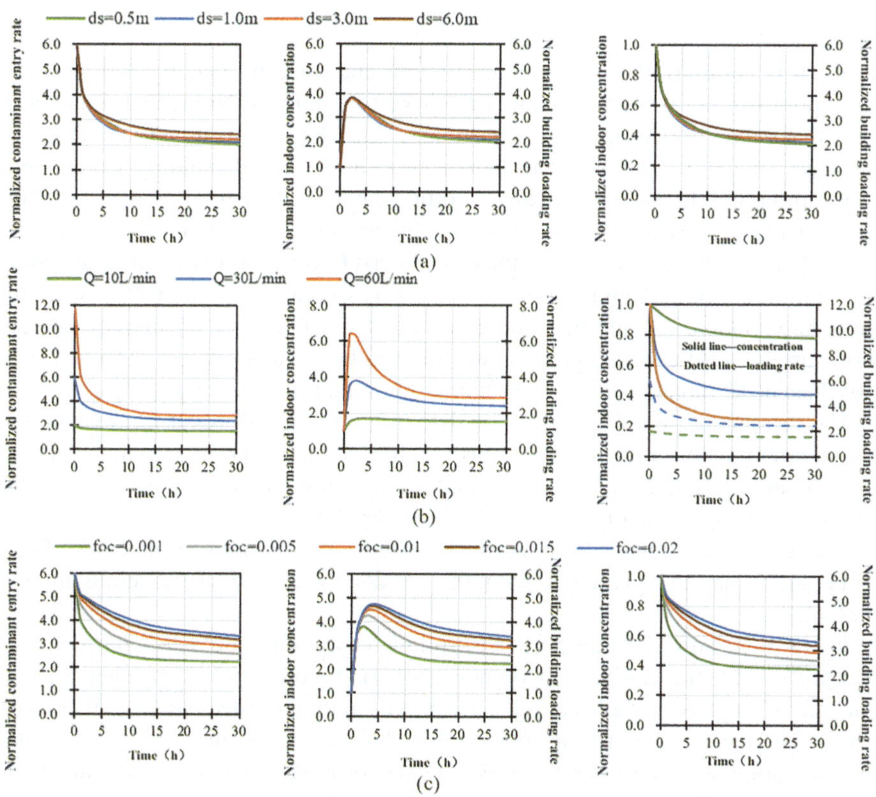

Fig. 5.8 Temporal variations with different vertical source-building separation distances (**a**), soil gas entry rate (**b**), and mass fractions of organic carbon (**c**).Column 1: temporal variations of the normalized contaminant entry. Column 2: temporal variations of the normalized indoor air concentration/building loading rate, assuming a constant indoor air exchange rate. Column 3: temporal variations of the normalized indoor air concentration/building loading rate, based on an indoor air exchange rate linear to depressurization. Reprinted with permission from [25]. Copyright 2020 Springer Nature

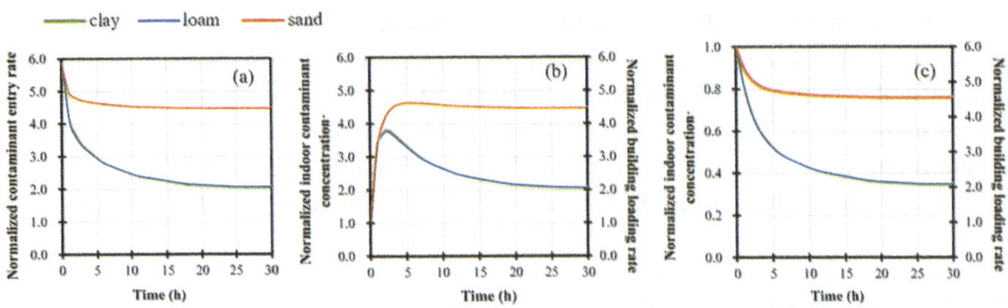

Fig. 5.9 Temporal variations of the normalized contaminant entry with different deep soil textures (**a**). Temporal variations of the normalized indoor air concentration/building loading rate assuming a constant indoor air exchange rate (**b**). Temporal variations of the normalized indoor air concentration/building loading rate based on an indoor air exchange rate linear to depressurization (**c**). Reprinted with permission from [25]. Copyright 2020 Springer Nature

(in the first column), the indoor concentration and building loading rate with two assumptions of the air exchange rate (in other two columns). Numerical simulations show that the entry rates increase immediately after the initialization of the BPC test and then gradually decrease. The indoor concentration and building loading rate share similar trends except that it takes a little longer for them to reach the peak if the air exchange rate is unchanged as shown in the middle column. By applying an unsteady air exchange rate, i.e., linear to the BPC operation strength, the indoor concentration would decline from the beginning due to the dilution [6], while the temporal variations in building loading rate are identical to those in emission rates, as shown in the third column. In all cases, the simulated values would drop to 1/3–1/2 of their peak values at the end, 30 h after the initiation of the test [13].

The strength of the BPC application determines indoor depressurization and thus the volumetric flow rate into the constructure. Actually, the maximum levels of contaminant entry rate, as well as the building loading rate, are almost linear to the indoor depressurization during the test. However, the semi-steady-state levels at the end of the test do not increase as much as the BPC strength in all cases. These observations are in agreement with the results by analytical modeling in Yao et al. [13].

In transient soil gas transport, adsorption/desorption should be considered as affected by the fraction of soil organic carbon. Figure 5.8c shows that a lower fraction of organic carbon can cause a more significant decrease in all VI risk indicators rate after reaching the peak. The reason is that the role of the subfoundation soil as a reservoir is significant for releasing more contaminant soil gas, if with a higher organic carbon fraction. Nevertheless, in general, the influences of organic carbon are not crucial in simulations [13].

On the other hand, the soil texture may play a much more important role in determining the performance of the BPC test. It is shown in Fig. 5.9 the simulated responses of three VI risk indicators for cases of sand, loam, and clay. It should be noted that the simulated results are identical in cases of loam and clay. The temporal trends of those variables are pretty similar regardless of the soil type. However, at the end of the BPC test, the simulated values can maintain at about 80% of their maximum in the sand case, while the corresponding figure is 40 or 50% for the other two [13].

5.4.3 Requirements for an Effective BPC Application

There are two possible explanations for the better performance of BPC in the sand case. First, the stronger advection may sweep the contaminant soil gas beyond the building foundation perimeter into the building under BPC conditions. Alternatively, the enhanced advection helps draw all contaminants from upward diffusion from the vapor source into the building. If under natural conditions, the blocking effect of the building foundation may induce a lateral soil gas transport, and only a portion of the contaminant from the upward diffusion can migrate into the building. The rest would escape into the atmosphere through the open ground surface surrounding the building. These two possibilities are both dependent on advection, which is limited in loam and clay soil [13].

Assume it is the first reason, and it is reasonable to conclude that the contaminant profile beyond the foundation is influenced by advection due to BPC. Contaminant profiles are shown after 0 and 12 h of indoor depressurization, respectively, for the sand soil in Fig. 5.10. It suggests that the near-foundation contaminant soil gas profiles are almost unchanged during the test. In fact, they are almost the same as that in cases with only diffusion. Such comparison indicates that the first explanation is

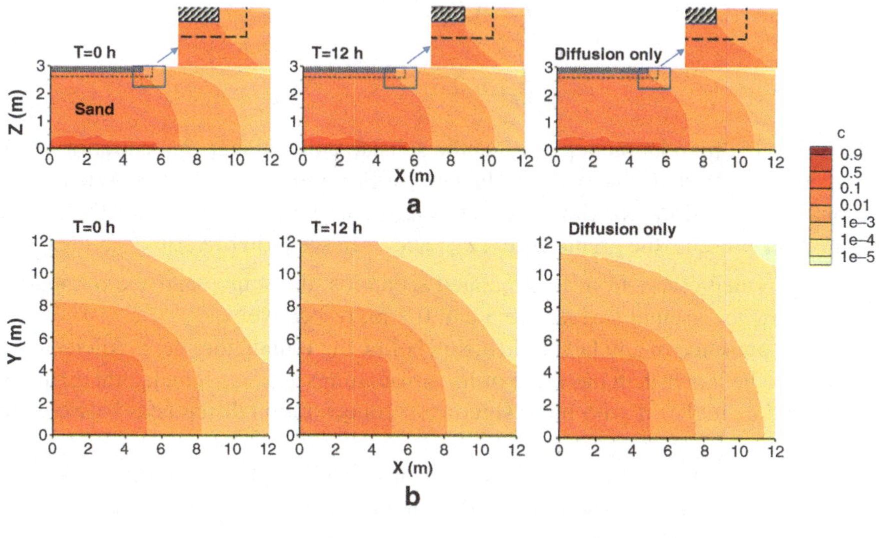

Fig. 5.10 Changes of normalized soil gas concentration profile after applying the BPC methods: vertical profile through the building and source footprint centerlines (**a**) and horizontal cross-section at the foundation depth (**b**). Reprinted with permission from [25]. Copyright 2020 Springer Nature

invalid, and the advection is not expected to play a role beyond the building perimeter [13].

So only one explanation is left and further examined with some simple equations. As the adsorption/desorption can be negligible based on the above discussion, a mass balance can be assumed in the subfoundation among the contaminant upward diffusive transport, inward and outward flow rate of the building [25]:

$$D_{\mathrm{eff}}^{\mathrm{T}} \frac{c_{\mathrm{s}}}{L_{\mathrm{T}}} A_{\mathrm{B}} = c_{\mathrm{ss}} Q_{\mathrm{soil}} + f(c_{\mathrm{ss}}) \tag{5.16}$$

where c_s is the source concentration [M/L^3], L_T is the thickness of the deep layer [L], A_B is the building footprint size [L^2], c_{ss} is the soil gas concentration of the subslab zone at the foundation depth [M/L^3], and $f(c_{ss})$ is the mass flow rate of contaminants escape part [M/T]. $D_{\mathrm{eff}}^{\mathrm{T}}$ is the total effective diffusivity through the deep soil layer [L^2/T], which can be calculated by [12].

$$\frac{D_{\mathrm{eff}}^{\mathrm{T}}}{L_{\mathrm{T}}} = \frac{1}{\int_0^{ds} \frac{1}{D_{\mathrm{eff}}(Z)} dz} \tag{5.17}$$

where ds is the source depth below the soil surface.

The contaminant concentration beyond the building perimeter at the foundation depth can be estimated by considering the blocking effect of the foundation negligible [13]:

$$\frac{\int_0^{df} \frac{1}{D_{\mathrm{eff}}(Z)} dz}{\int_0^{ds} \frac{1}{D_{\mathrm{eff}}(Z)} dz} = \frac{c_{\mathrm{sg}}}{c_{\mathrm{s}}} \tag{5.18}$$

where df is the foundation depth, c_{sg} is the subslab contaminant concentration at the foundation depth and out of the influence by the foundation [M/L^3]. When $f(c_{ss})$ becomes zero, and the maximum value of c_{ss} could be estimated using Eq. (5.16). Figure 5.11 presents the predicted c_{ss}, c_{sg}, and $\frac{c_{ss}}{c_{sg}}$ in scenario of various soil textures without any depressuriztion. In simulated scenarios, the source and the foundation depth below the ground surface are 3 and 0.2 m, respectively.

For the presence of any lateral diffusion below the foundation, the predicted c_{ss} is expected to be larger than the value of its surroundings, i.e., c_{sg}. The lateral diffusion rate should be higher if with more significant concentration differences between the two [13].

Figure 5.11 shows that the soil texture or actually soil particle diameter has a significant effect on determining $\frac{C_{ss}}{C_{sg}}$. The lateral diffusion is expected with coarser soils, i.e., the mean particle diameter is larger than 0.3 mm. It should be noted that lateral diffusion occurs when the ratio is larger than 1. For example, $\frac{C_{ss}}{C_{sg}}$ can be as

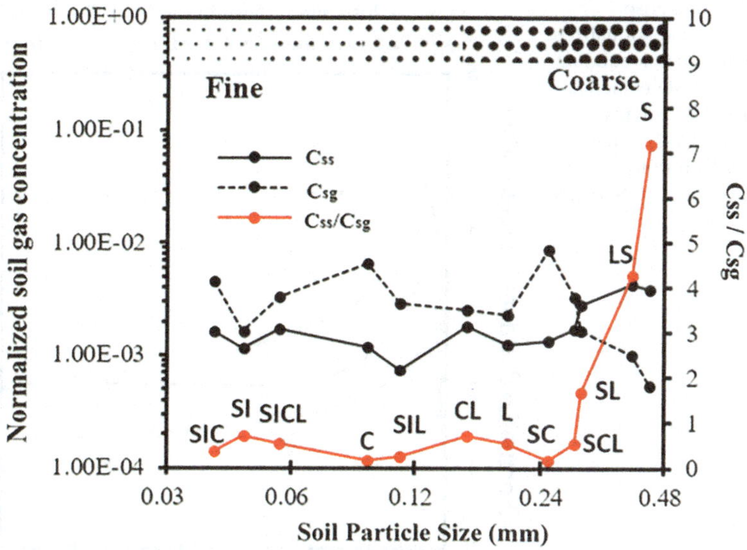

Fig. 5.11 Influence of the mean deep soil particle diameter on the normalized soil gas concentration of/near the subslab zone at the foundation depth, calculated by 1D models. Reprinted with permission from [25]. Copyright 2020 Springer Nature

high as 7 in the sand scenario, suggesting the building loading rate can be increased by about six times, generally consistent with the 4× to 5× increase in the above simulations [13]. Figure 5.12 compares 3D and 1D numerical simulations. It suggests the transient building loading rates in both cases could be similar if with a lower soil permeability. Such results justified the previous conclusion that the limit of a 1D model in simulation only vertical transport can not capture the whole picture in such a scenario with significant lateral transport [13].

Figure 5.13 summarizes the field tests of BPC. Due to the limited data availability, some slight differences can be observed in the estimations, as explained in the figure caption. The summary shows the normalized building loading rate could be expected to be more than four if with a preferential pathway. For example, a tunnel for utility had been discovered as an alternative VI pathway at the site of Moffett Field with an increase of four [6, 27]. A higher observation might suggest the potential existence of a preferential VI pathway, such as 10 at Hill AFB Residence #1 [24]. For the ASU House, the increase of DCE was observed to be higher than 300 times, with the identification of a land drain for the preferential VI pathway [2, 6, 8].

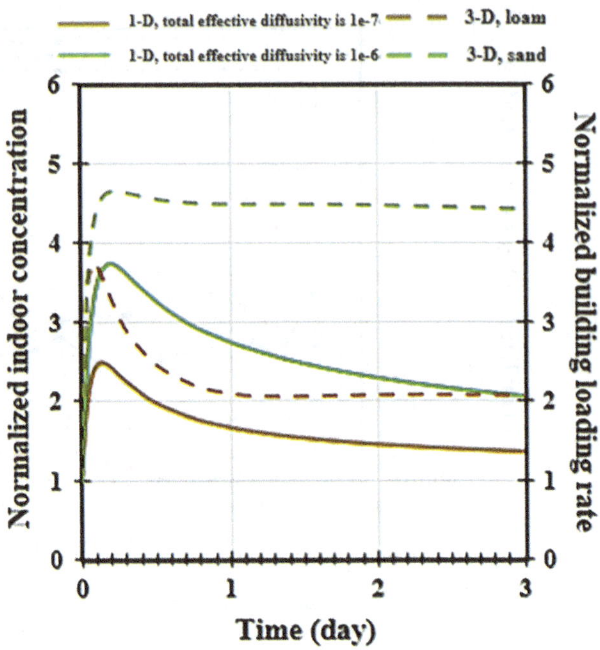

Fig. 5.12 Comparison between predicted normalized indoor concentration/building loading rate of 3D numerical models and 1D analytical models [13, 25]. Reprinted with permission from [25]. Copyright 2020 Springer Nature

5.5 The Application of Building Pressure Cycling to Generate Aerobic Barrier in Petroleum Vapor Intrusion

Currently, the majority of BPC tests have been performed mainly at chlorinated chemical sites, while just one or two involve petroleum products. Even in the latter, the indoor VOCs were later identified from background vapor sources [6, 7]. In petroleum vapor intrusion (PVI) risk assessments, US EPA [28] suggests employing oxygen-limited degradation, which would attenuate vapor concentration during soil gas transport. Such biodegradation can be enhanced or prevented by adjusting the availability of oxygen in the soil. Luo et al. [29] conducted a proof-of-concept study to generate a subsurface aerobic barrier at a petroleum site by injecting fresh air into the soil. In theory, there is a possibility of achieving similar effects by using BPC to induce pressurization in the building.

Liu et al. [30] used the previously developed Brown model to study the temporal variations in building loading rates and soil gas profiles during the BPC tests at PVI sites. The numerical simulations were performed to examine cases with different source depths and strengths and reaction rate constants by manipulating indoor-outdoor pressure differences of negative or positive values.

Figure 5.14 compares simulated contaminant profiles of oxygen and benzene with field observations [29]. Numerical simulation was conducted basically with the field conditions, but the contaminants were represented by benzene with a source vapor concentration of 150 mg/L at 10 m below the ground surface. Oxygen only

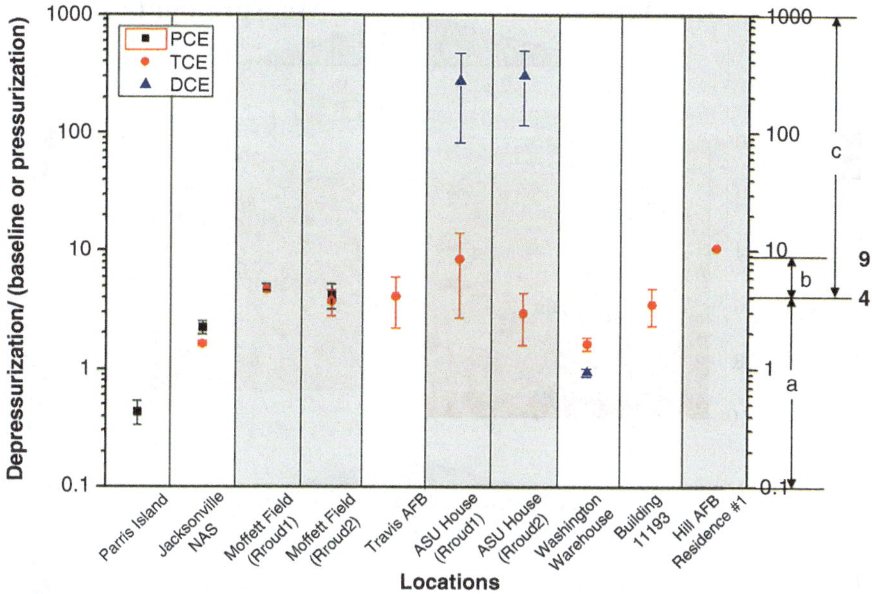

a. Traditional soil VI pathway
b. Traditional soil VI pathway with a strong source
c. Preferential pathway

Fig. 5.13 Field data. Grey areas refer to the sites with preferential pathways reported in (or speculate by) the references [2, 5–8, 24, 27]. Reprinted with permission from [25]. Copyright 2020 Springer Nature

exists in the shallow soil (i.e., depth <1.5 m) surrounding the building foundation but is almost depleted beneath it, as the high-moisture silt soil virtually isolated the subfoundation soil environment from the atmosphere. The hydrocarbon concentration is horizontally uniform for both simulation and observations, suggesting the low-permeability soil plays a similar role in blocking soil gas diffusion as the building foundation slab. As the limited availability of oxygen in the subfoundation, the concentration attenuation is also insignificant. In fact, soil gas concentrations lower than 60 mg/L were only observed in the aerobic region surrounding the building foundation.

Then the model was used to examine the temporal behaviors of building loading rate and indoor concentrations in 60-day indoor depressurization for chlorinated VI (CVI) and PVI. In short, the responses are identical regardless of the chemical type. This is good news for site investigators when trying to obtain a maximum reasonable exposure with indoor air sampling by using BPC to minimize temporal and spatial variations.

To examine the potential of BPC application in improving aerobic conditions in the subfoundation, the duration of indoor pressurization was extended to 60 days to provide a sustainable oxygen supply. Figure 5.15 shows the responses of normalized

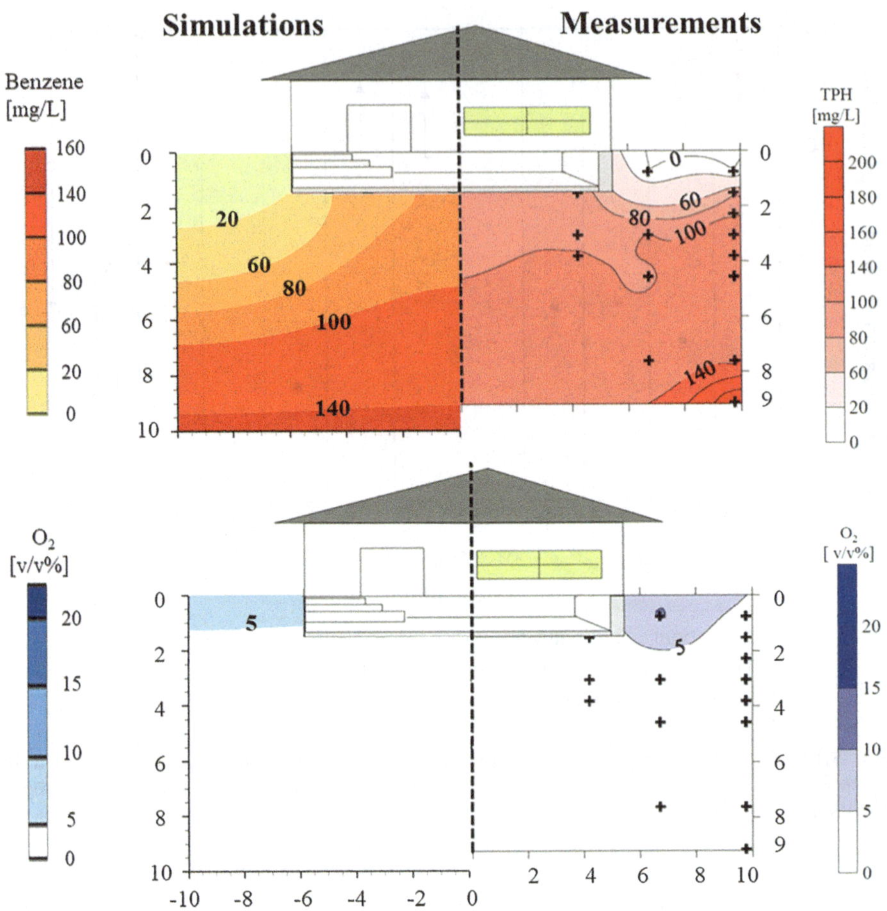

Fig. 5.14 Comparisons between simulations and field measurements with baseline conditions. The simulations were reprinted with permission from [25]. Copyright 2020 Elsevier B.V. The field observations were reprinted with permission from [29]. Copyright 2013 American Chemical Society

building loading rates in PVI and CVI during the test. It suggests that building loading rates are identical for both cases in a relatively short period (i.e., 1 day) after the test. After dropping to about 0.01, the decay of the building loading rate in CVI becomes much slower while that in PVI continues. At the end of the test (i.e., $t = 60$ days), the building loading rate in PVI is lower than that in CVI by three orders of magnitude, indicating the influence of biodegradation [25].

Figure 5.16 shows calculated and measured concentration profiles of oxygen and contaminant at a PVI site after fresh air injection into the soil [29, 30]. In the simulations, the fresh air enters the soil through foundation cracks, while in the field experiment it is through a subslab pipe system. Figure 5.16a shows the transient contaminant and oxygen concentration profiles during the first 30 days after the test

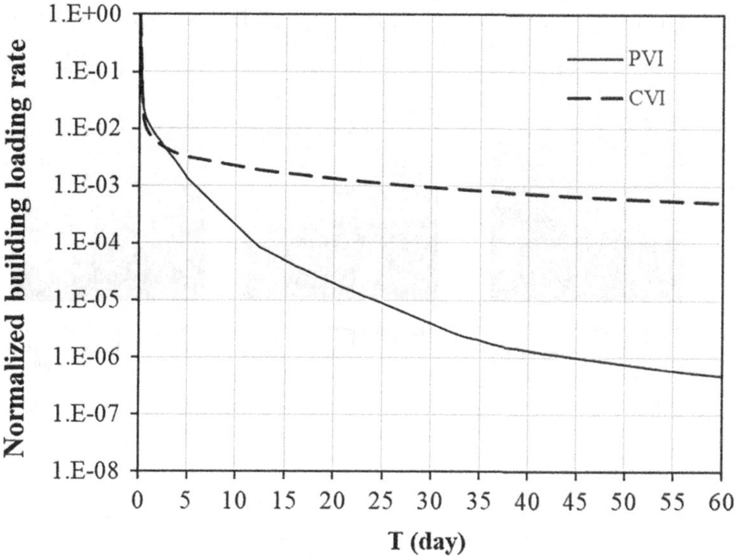

Fig. 5.15 Temporal variations of normalized building loading rate with reaction rate constants after the 60-day application of BPC pressurization

initiation. It suggests the aerobic air sweep zone (>1% v/v) below the foundation is insignificant after a day, and the subslab benzene concentrations drop to about 1/3 of that under baseline conditions shown in Fig. 5.14. The decrease is due to the dilution effect of fresh air from the foundation crack, consistent with the results reported in Fig. 5.15 [25].

The results in Fig. 5.16a also indicate that the significant biodegradation can only be observed in a long-term test. With sustainable indoor pressurization, the aerobic air sweep zone gradually expands to the whole subfoundation zone, where the contaminant concentrations become lower than 20 mg/L and 0.1 mg/L in 10 and 30 days, respectively. It shows in the first 10 days, the aerobic conditions could be formed in most subfoundation areas, but the significant decrease of hydrocarbon soil gas concentrations caused by biodegradation would not occur until a few days later. Moreover, the simulated soil gas concentrations profiles still do not reach steady-state after 30-day BPC application [25].

Figure 5.16b presents the comparison with field measurement at $t = 60$ days. In both simulations and observations, the oxygen can migrate as deep as 5–6 m bgs into the soil. Comparatively, the simulated aerobic air sweep zone is a little smaller than the field measurement, possibly because, in the field test, the horizontal well is installed at 1.5 m below the foundation, while in the simulated scenario, the air injection is through the foundation cracks. For the hydrocarbon soil gas concentration profile, the observed iso-concentration line of zero concentration is located about 3 m bgs, and correspondingly the simulated subfoundation concentration is

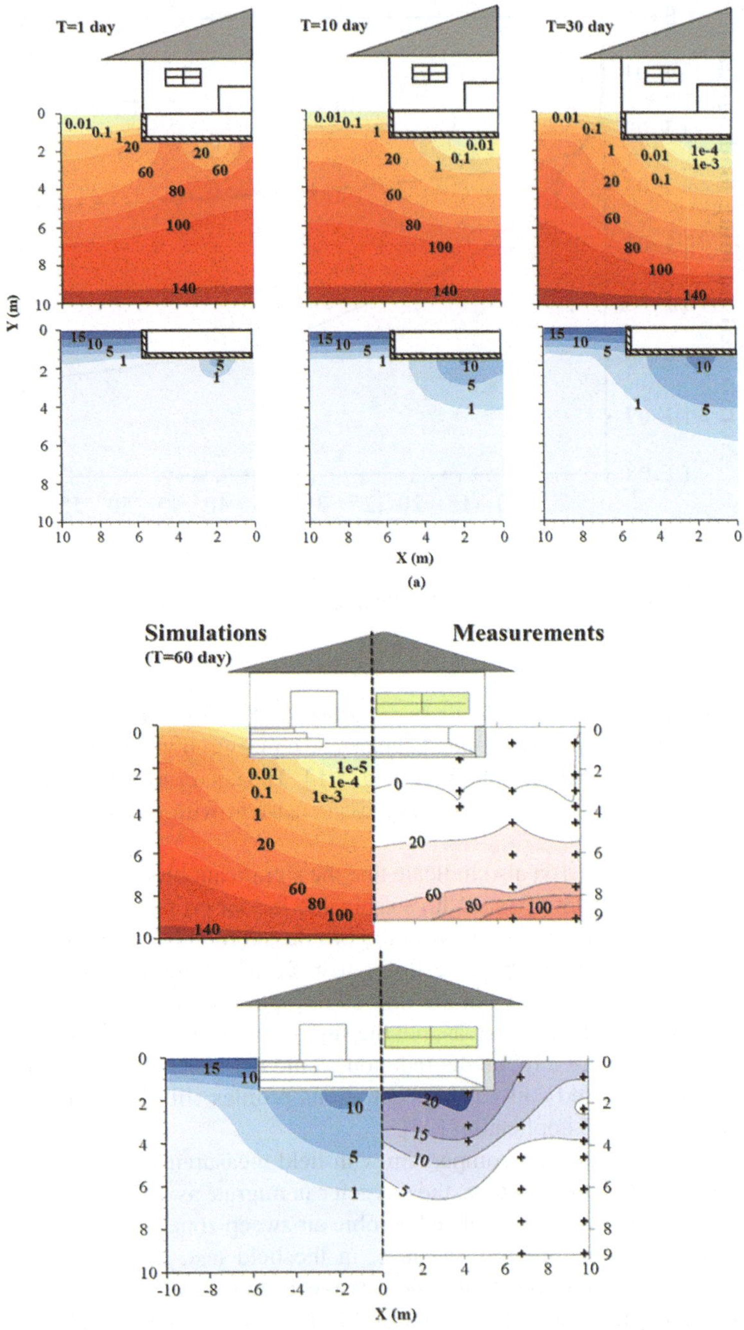

Fig. 5.16 (**a**) Temporal variations of simulated soil gas profiles in the first 30 days and (**b**) the comparison with field observations at 60 days. The simulations were reprinted with permission from [25]. Copyright 2020 Elsevier B.V., and the field observations were reprinted with permission from [29]. Copyright 2013 American Chemical Society

generally less than 0.01 mg/L. At 60 days after the BPC initiation, the simulated concentrations profiles achieve a pseudo-steady-state [25].

References

1. US EPA. Fluctuation of indoor radon and VOC concentrations due to seasonal variations. Washington DC: Office of Research and Development. EPA 600-R-12-673.; 2012.
2. Guo YM, Holton C, Luo H, Dahlen P, Gorder K, Dettenmaier E, Johnson PC. Identification of alternative vapor intrusion pathways using controlled pressure testing, soil gas monitoring, and screening model calculations. Environ Sci Technol. 2015;49(22):13472–82.
3. Pennell KG, Scammell MK, McClean MD, Ames J, Weldon B, Friguglietti L, Suuberg EM, Shen R, Indeglia PA, Heiger-Bernays WJ. Sewer gas: an indoor air source of PCE to consider during vapor intrusion investigations. Ground Water Monit Remediat. 2013;33(3):119–26.
4. Riis CE, Christensen AG, Hansen MH, Husum H, Terkelsen M. Vapor intrusion through sewer systems: migration pathways of chlorinated solvents from groundwater to indoor air. In: Proceedings of the 7th international conference on remediation of chlorinated and recalcitrant compounds, Monterey, CA. 24–27 May 2010. Madison, WI: Battelle Memorial Inst. 2010.
5. McHugh TE, Nickels TN. Final report: detailed field investigation of vapor intrusion processes. ESTCP project ER-0423. Houston, TX: GSI Environmental; 2008.
6. McHugh TE, Beckley L, Bailey D, Gorder K, Dettenmaier E, Rivera-Duarte I, Brock S, MacGregor IC. Evaluation of vapor intrusion using controlled building pressure. Environ Sci Technol. 2012;46(9):4792–9.
7. Beckley L, Gorder K, Dettenmaier E, Rivera-Duarte I, McHugh T. On-site gas chromatography/mass spectrometry (GC/MS) analysis to streamline vapor intrusion investigations. Environ Forensics. 2014;15(3):234–43.
8. Holton C, Guo YM, Luo H, Dahlen P, Gorder K, Dettenmaier E, Johnson PC. Long-term evaluation of the controlled pressure method for assessment of the vapor intrusion pathway. Environ Sci Technol. 2015;49(4):2091–8.
9. Guo YM, Dahlen P, Johnson PC. Development and validation of a controlled pressure method test protocol for vapor intrusion pathway assessment. Environ Sci Technol. 2020;54(12): 7117–25.
10. Yao Y, Mao F, Ma S, Suuberg EM, Tang X. Three-dimensional simulation of land drains as a preferential pathway for vapor intrusion into buildings. J Environ Qual. 2017;46(6):1424–33.
11. Holton C, Luo H, Dahlen P, Gorder K, Dettenmaier E, Johnson PC. Temporal variability of indoor air concentrations under natural conditions in a house overlying a dilute chlorinated solvent groundwater plume. Environ Sci Technol. 2013;47(23):13347–54.
12. Johnson PC, Ettinger RA. Heuristic model for predicting the intrusion rate of contaminant vapors into buildings. Environ Sci Technol. 1991;25(8):1445–52.
13. Yao Y, Zuo J, Luo J, Chen Q, Strom J, Suuberg EM. An examination of the building pressure cycling technique as a tool in vapor intrusion investigations with analytical simulations. J Hazard Mater. 2020;389:121915.
14. Millington RJ, Quirk JP. Permeability of porous solids. Trans Faraday Soc. 1961;57:1200–7.
15. Abreu LDV, Johnson PC. Effect of vapor source – building separation and building construction on soil vapor intrusion as studied with a three-dimensional numerical model. Environ Sci Technol. 2005;39(12):4550–61.
16. Ström JGV, Guo Y, Yao Y, Suuberg EM. Factors affecting temporal variations in vapor intrusion-induced indoor air contaminant concentrations. Build Environ. 2019;161:106196.
17. Pitts DR, Sissom LE. Schaum's outline of theory and problems of heat transfer. New York: McGraw-Hill; 1997.
18. Mills AF, Coimbra CFM. Basic heat and mass transfer. San Diego: Temporal Publishing; 2015.
19. Baehr HD, Stephan K. Heat and mass transfer. New York: Springer; 2006.

20. Shen R, Pennell KG, Suuberg EM. Influence of soil moisture on soil gas vapor concentration for vapor intrusion. Environ Eng Sci. 2013;30(10):628–37.
21. Yao Y, Wang Y, Zhong Z, Tang M, Suuberg EM. Investigating the role of soil texture in vapor intrusion from groundwater sources. J Environ Qual. 2017;46(4):776–84.
22. Pennell KG, Bozkurt O, Suuberg EM. Development and application of a three-dimensional finite element vapor intrusion model. J Air Waste Manage Assoc. 2009;59(4):447–60.
23. Liu Y, Man J, Wang Y, Xiao Y, Tang W, Chen Q, Yao Y. Numerical study of the building pressure cycling method for evaluating vapor intrusion from groundwater contamination. Environ Sci Pollut Res Int. 2020;27(28):35416–27.
24. Dawson H, Wertz W, McAlary T, Gabris T, Carr D. Assessing building susceptibility to vapor intrusion with building pressure cycling. In: Poster presented at SERDP & ESTCP symposium, Washington, DC. 2018.
25. Liu Y, Man J, Wang Y, Xiao Y, Tang W, Chen Q, Yao Y. Numerical study of the building pressure cycling method for evaluating vapor intrusion from groundwater contamination. Environ Sci Pollut Res. 2020;27(28):35416–27.
26. McAlary TA, Wertz W, Mali DJ. Demonstration/validation of more cost-effective methods for mitigating radon and VOC subsurface vapor intrusion to indoor air. Boca Raton: Geosyntec Consultants, Inc; 2018.
27. McHugh TE, Loll P, Eklund B. Recent advances in vapor intrusion site investigations. J Environ Manag. 2017;204:783–92.
28. US EPA. Technical guide for addressing petroleum vapor intrusion at leaking underground storage tank sites. Washington DC: Office of Underground Storage Tanks. EPA 510-R-15-001.; 2015.
29. Luo H, Dahlen PR, Johnson PC, Peargin T. Proof-of-concept study of an aerobic vapor migration barrier beneath a building at a petroleum hydrocarbon-impacted site. Environ Sci Technol. 2013;47(4):1977–84.
30. Liu Y, Verginelli I, Yao Y. Numerical study of building pressure cycling to generate sub-foundation aerobic barrier for mitigating petroleum vapor intrusion. Sci Total Environ. 2021;779:146460.

6 Vapor Intrusion Risk Assessments in Brownfield Redevelopment

The concept of a brownfield site is introduced in the redevelopment and utilization of land, which is often more complex than other site development processes due to the presence of objective or intended environmental contamination. The US "Brownfield Act" (The Small Business Liability Relief and Brownfield Revitalization Act) definition: The term "brownfield site" means real property, the expansion, redevelopment, or reuse of which may be complicated by the presence or potential presence of a hazardous substance, pollutant, or contaminant. The concept of a brownfield site in the UK emphasizes that it was once utilized but is underutilized in its current state, without regard to whether the cause of that current state is contamination. The Canadian and European CABERNET (Concerted Action on Economic and Brownfield Regeneration Network) brownfield concept is similar to that of the United States. In Germany, brownfields include land and underutilized buildings; in France, agricultural land is included in the scope of brownfields [1]. However, contamination and the need for redevelopment and use are common to the brownfield concept. China does not have brownfields within the scope of official documents. Nonetheless, the primary connotation of contaminated land in China is similar to brownfields, and its definition mainly emphasizes the inclusion of contamination, but the actual management process is mainly managed in the relevant aspects of land redevelopment.

After the land is contaminated, some method is needed to evaluate the possible hazard level of the contamination, which is used to determine whether the land is required for subsequent management. The risk assessment method is commonly used internationally to do the above. Risk assessment of soil/groundwater contamination refers to the characterization of the likelihood of the occurrence of inevitable harmful consequences caused by soil/groundwater contamination using a probabilistic approach. Contamination risks are usually classified into health risks and ecological risks. In practice, the operation also focuses on the contamination risks to other environmental media and thus indirectly assesses health risks and ecological risks. The health risk is the likelihood of injury, illness, or death from human exposure to a contaminated environment. Ecological risk is the probability or

Y. Yao, Q. Chen, *Vapor Intrusion Simulations and Risk Assessments*,
https://doi.org/10.1007/978-981-19-2700-3_6

likelihood that a soil contaminant will cause damage to some element of the ecosystem or the ecosystem itself.

The risk assessment of brownfield contamination is the same as that of other land contamination in terms of basic concepts and methods. However, since brownfield sites will be subsequently redeveloped, their land-use status will change from the current situation, which means that it is more difficult to obtain substantial evidence of risk in brownfield risk assessment and relies more on model prediction. Third, brownfield redevelopment is often located in large-scale urban development zones. The subsequent land reuse is usually based on the use of artificial construction facilities, which is more concerned about protecting health risks and involve relatively few ecological risks. Even if it does, it is mainly. Finally, as redevelopment and construction are required later, the protection of existing facilities on brownfield sites is less demanding, which means adequate intrusive investigations can usually be conducted. More adequate engineering treatments can be adopted subsequently.

Human health risk assessment is essential for environmental management and treatment of soil and groundwater contamination at contaminated sites. It has been widely used in contaminated site management and development at home and abroad. The criteria for distinguishing the contamination level of contaminated sites and subsequent treatment plans are determined mainly by a human health risk assessment of contaminated soil and groundwater in contaminated sites. At present, in the process of the comprehensive assessment of the hazards of contaminated sites, that is, the so-called health risks, different foreign organizations or groups have developed several exposure evaluation models for health risk assessment according to different needs, such as CLEA, RISC, CSOIL, RBCA, ROME, Sniffer, RISC-Human, and so on. Among them, RBCA, CLEA, and CSOIL models are more widely used. Regardless of the exposure assessment models, vapor intrusion, i.e., volatile organic compounds (VOCs) from groundwater or soil, is an integral part of the exposure pathway through vapor intrusion into the room and inhalation by humans.

For brownfield sites contaminated with volatile pollutants, especially VOCs, the most critical contamination exposure pathway that poses a health risk is vapor intrusion, considering that the construction of aboveground facilities during future brownfield reuse often prevents direct soil/groundwater exposure. It is defined as the process by which volatile contaminants are released from contaminated underlying soil or groundwater, migrate through the soil, and eventually enter the interior through cracks in the foundations of surface buildings.

6.1 The Basic Concept and Process of Vapor Intrusion

Vapour intrusion is a standard term which describes volatile chemicals from underground contaminated soil and groundwater migrating through cracks in building foundations and into aboveground buildings, usually in scenarios where soil or groundwater contains volatile contaminants underneath and a house overlying. Volatile contaminants vaporize from the soil or groundwater that is the source of contamination, migrate into the soil, and then infiltrate into the building like radon intrusion, ultimately posing a health risk to people living and working in the building [2]. One of the concerns of vapor intrusion is the concentration attenuation during the vapor migration process, which mainly occurs during the migration of vapor in soil, such as the diffusion, convection, adsorption, and possible degradation of vapor, in addition to the dilution effect of indoor air when entering the building. Defined as the ratio of indoor air concentration to underground concentration, the term "attenuation factor" is used as an overall measure of reduction in vapor concentration over space and time during gas-phase transport [3].

These volatile pollutants include VOCs, semi-volatile organic compounds (SVOCs), and some other inorganic compounds such as mercury (Hg) and hydrogen sulfide (H_2S). Typically, the most common chemical sources of pollutants are VOCs, which include tetrachloroethylene(PCE), trichloroethylene (TCE), vinyl chloride (VC), carbon tetrachloride (CTC), naphthalene, benzene, toluene, ethylbenzene, and xylenes (BTEX). Such VOCs generally pose a chronic health risk to individuals living in buildings through the inhalation of indoor air, with the potential for chronic risk from their volatile degradation products through vapor intrusion. In addition, some volatile pesticides, such as chloromon, Aldrin, and Lindane, also pose vapor intrusion risks. Some less volatile substances, such as PCBs or mercury, also pose a risk.

VOCs are organic compounds or mixtures of them that can be volatilized or vaporized at ambient temperature and pressure. Generally, VOCs have high vapor pressures and Henry's constants, low or medium solubilities and low molecular weights. Quantitatively, a substance with a molecular weight less than 200 g/mol, a vapor pressure greater than 1 mmHg, or a Henry's constant greater than 10^{-5} atm-m^3/mol is generally considered volatile.

The process of vapor release from subsurface sources to migration within the soil is often governed by a complex set of factors, and many variables can influence vapor intrusion [4]. These variables include, but are not limited to, the following: contaminant concentration, source depth, groundwater burial depth, soil type and spatial variation, biomass, building materials, and construction conditions, seasonal variations in climate, barometric pressure, permafrost, and building habits. Vapors migrate from higher concentration areas to lower concentration areas under the influence of diffusion in relatively deep soils and from areas of higher pressure to areas of lower pressure under the action of air currents in soils near the surface or below buildings. Once the vapors reach the area below the building, they can accumulate until they migrate upward into the building. The pressure difference between the soil and the interior of the building drives the vapor from the soil through the building gaps and into the building. This pressure difference is often called the chimney effect and can be influenced by heating systems, wind, and basement structures.

A complete vapor intrusion model should include the entire process from the contamination source to the contaminated receptor, including the vapor migration within the soil and into the building [5]. In some cases, simplified models include only the former processes because the most critical environmental factors, such as convection, diffusion, adsorption, and biodegradation, all occur in this process. All vapor intrusion models, whether complete or simplified, have difficulty considering the effects of all relevant environmental factors, and thus the better vapor intrusion models focus on the critical environmental factors in the current scenario.

For soil vapor transport processes, a quantitative description of the contaminants leaving the source to reach the vicinity of the building of concern is necessary. For a typical vapor intrusion scenario, i.e., a groundwater source and an aboveground building of concern, the contaminant concentration distribution is determined primarily by diffusion and, in some cases, by convection, adsorption, and biodegradation.

Although convection as a whole is a minor factor in the diffusion of pollutants, it must be taken into account in local areas (e.g., areas of high-pressure gradients at building boundaries). Usually, we assume that the flow of gases within the soil obeys Darcy's law, which means that the magnitude and direction of their movement depend on the pressure gradient in the soil. When this pressure gradient is not very large relative to the atmospheric pressure, the gas can be assumed to be incompressible. This is an almost universal assumption for most vapor intrusion scenarios.

Generally speaking, the instantaneous form of Darcy's law is

$$\frac{\partial p}{\partial t} - \frac{p_{\mathrm{atm}} k_{\mathrm{g}}}{\varphi_{\mathrm{g}} \mu_{\mathrm{g}}} \nabla \cdot (\nabla p) = 0 \tag{6.1}$$

$$q_{\mathrm{g}} = \frac{k_{\mathrm{g}}}{\mu_{\mathrm{g}}} \nabla p \tag{6.2}$$

where: p is the inter-soil gas pressure; t is the time; p_{atm} is the atmospheric pressure; k_{g} is the coefficient of permeability of the soil to the gas; φ_{g} is the soil porosity; μ_{g} is the gas viscosity coefficient; q_{g} is the flux per unit area.

This equation ignores the effects of density change and gravity, both of which, as stated above, have a negligible effect on this problem (vapor migration in the soil).

The transport of contaminants in soil is usually described as a convective diffusion equation that

$$\varphi_{\mathrm{g,w,s}} \frac{\partial c_{\mathrm{ig}}}{\partial t} = -\nabla \cdot (q_{\mathrm{g}} c_{\mathrm{ig}}) - \nabla \cdot \left(\frac{c_{\mathrm{ig}}}{H_{\mathrm{i}}} q_{\mathrm{w}}\right) + \nabla \cdot (D_{\mathrm{i}} \nabla c_{\mathrm{ig}}) - R_{\mathrm{i}} \tag{6.3}$$

$$\varphi_{\mathrm{g,w,s}} = \varphi_{\mathrm{g}} + \frac{\varphi_{\mathrm{w}}}{H_{\mathrm{i}}} + \frac{k_{\mathrm{oc,i}} f_{\mathrm{oc}} \rho_{\mathrm{b}}}{H_{\mathrm{i}}} \tag{6.4}$$

where: φ_{g} is the water-filled porosity of the soil (water volume in the soil to total soil volume ratio); H_{i} is Henry's constant for pollutant i (the ratio of pollutant gas-phase concentration to liquid-phase concentration); $k_{\mathrm{oc,\ i}}$ is the organic carbon sorption coefficient of pollutant i in the soil; f_{oc} is the mass fraction of organic carbon in the soil; ρ_{b} is the soil density; c_{ig} is the gas-phase concentration of pollutant i; q_{w} is the water flow in the seepage zone (if present); D_{i} is the total effective diffusion coefficient of contaminant migration in porous media; R_i is the biodegradation rate of pollutant i; $\varphi_{g,\ w,\ s}$ indicates the total effective porosity.

$\varphi_{g,\ w,\ s}(\partial c_{ig}/\partial t)$ The time dependence of contaminant masses in soil gas, water, and organic carbon is described. $-\nabla \cdot (q_g c_{ig}) - \nabla \cdot ((c_{ig}/H_i)q_w)$ The transport of contaminants by convection in the presence of groundwater or gas flow is described, and the $\nabla \cdot (D_i \nabla c_{ig})$ vapor-phase diffusion of contaminants is described (liquid-phase diffusion is negligible because it is much smaller than gas-phase diffusion). This equation also describes the transport of oxygen in the soil. This will be used in the biodegradation model under aerobic conditions [6].

In most cases, the pollutant concentrations in the vapor intrusion scenarios are low enough that the uncoupled solution of the above two equations does not cause significant errors. The diffusion coefficients in D_i obtained through the Millington-Quirk criterion [7] as

$$D_i = D_i^g \frac{\varphi_g^{10/3}}{\varphi_t^2} + D_i^w \frac{\varphi_w^{10/3}}{\varphi_t^2} \tag{6.5}$$

where: D_i^g is the diffusion coefficient of the pollutant in the air; D_i^w is the diffusion coefficient of the pollutant in water; φ_t is the total porosity.

Vapor intrusion models are usually divided into two types, one is a simple screening model, mainly used for site assessment, and most use a one-dimensional analytical model, and the other is more detailed [8]. The multidimensional numerical model considers most of the influencing factors of the decay concentration of vapor-phase pollutants. The former has a more straightforward computational process due to the simplification of scenarios, but the accuracy of predictions may not be high [9], while the latter can simulate complex environmental processes but requires a lot of computational work [10]. However, the difference between the two is not absolute. Overall, the Johnson and Ettinger model (the J-E model) is the most widely used and representative vapor intrusion model at present due to its successive adoption by EPA, EA, and the Chinese Ministry of Environmental Protection as the primary recommended model for vapor intrusion screening or risk assessment; followed by the Volasoil model used in the Dutch risk assessment.

6.2 The Basic Theory of the J-E Model

The J-E model was proposed by Paul Johnson and Robert Ettinger, chemical engineers at Shell, in 1991 to simulate vapor intrusion processes [11]. It incorporates the processes of contaminant transport and transfers through both soil water and soil gas media into the analysis. This type of model is aimed at the volatilization process of pollutants present in the soil in the indoor environment is divided into two primary migration stages: the first stage is the transport process from the pollutant source to the soil around the bottom of the building; the second stage is the change process from the movement of soil around the building to the indoor environment (Fig. 6.1).

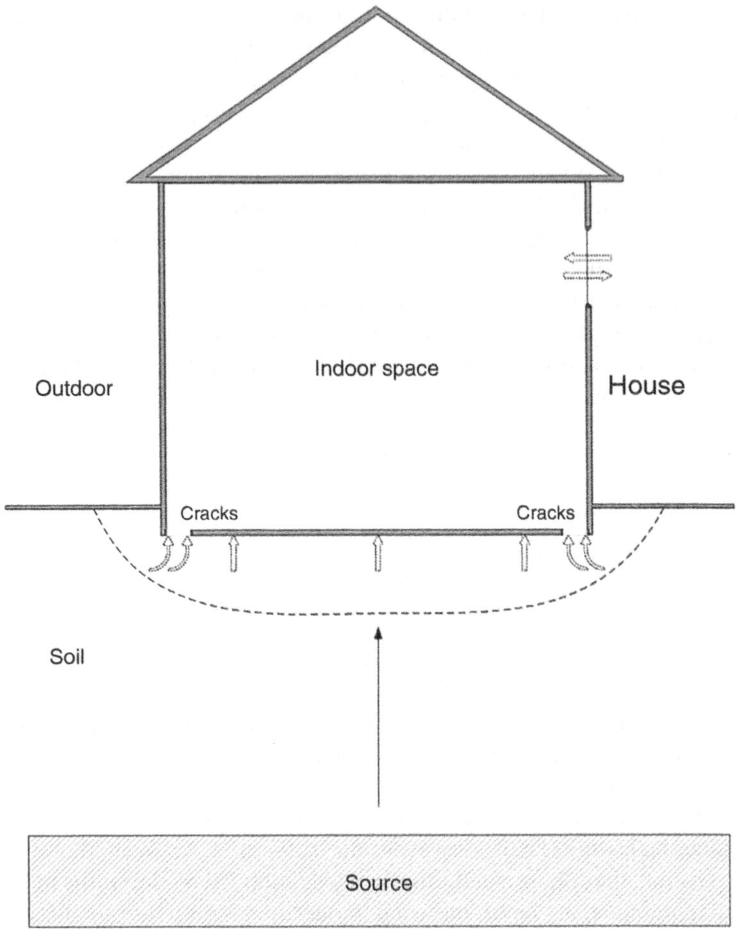

Fig. 6.1 J-E model

The J-E model is applied to the analysis of indoor volatilization processes of pollutants, and the calculation process implemented for the calculation of pollution effects is depicted in Fig. 6.2. In the actual evaluation, the gas-phase concentration of soil at the base of the building is used as an unknown quantity in the calculation of the model, and the value of this indicator is determined by the association of equations. Therefore, the J-E model for the indoor pollutant volatilization process is solved and calculated by establishing the equations for each migration process and then solving them in conjunction.

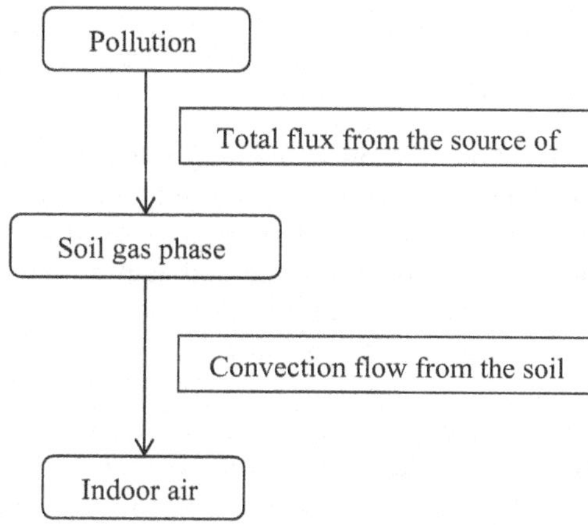

Fig. 6.2 the J-E model calculation flow chart

6.2.1 Fluxes from the Source to the Soil Near the Building Calculation

As mentioned above, the contaminant transfer process from the source to the soil near the building is mainly influenced by diffusion. In the steady case, assuming that the effects of convection, adsorption, and biodegradation are neglected, assuming that the soil properties are homogeneous, and considering only the variation of the concentration delayed vertical gradient (one-dimensional), the convective diffusion equation described in Sect. 6.1 is simplified as: $\varphi_{g,\ w,\ s}(\partial c_{ig}/\partial t) = \nabla \cdot (D_i \nabla c_{ig})$, which is integrated over the vertical length from the source to the building. Integrating, the following relationship can be obtained.

$$J_i = D_i \frac{C_{isa} - C_{isoil}}{L_s} \tag{6.6}$$

where J_i is the total flux from the source to the direction of the building; C_{isa} is the concentration of the pollutant at the source; C_{isoil} is the concentration of contamination in the soil area near the building; L_s is the distance from the source of contamination to the soil near the building.

In the case of heterogeneous soils, the total effective diffusion coefficient D_i is calculated by considering the liquid and gas phases and the influence of soil stratification.

6.2.2 Calculation of the Flux from the Soil Around the Base of the Building to the Interior of the Building

As mentioned above, the process of transport from the soil around the base of the building to the interior of the building requires a combination of convective and diffusive effects, and the following relationships can be obtained for the stable case.

$$E = Q_{\text{soil}} C_{\text{isoil}} - \frac{Q_{\text{soil}}(C_{\text{ia}} - C_{\text{isoil}})}{1 - \exp\left(\frac{Q_{\text{soil}} L_{\text{f}}}{D_{\text{crack}} A_{\text{crack}}}\right)} \tag{6.7}$$

where E is the flow of pollutants transported from the soil around the base of the building to the interior of the building; A_{crack} is the area of the building foundation fissures; C_{ia} is the indoor volatile exposure concentration; D_{crack} is the effective diffusion coefficient of the pollutant in the fissure; Q_{soil} is the total amount of air from the bottom soil to the chamber; L_{f} is the thickness of the base plate.

For Eqs. (6.10) and (6.11), the gas-phase concentration of the contaminant present in the soil around the bottom of the building C_{isoil} is an unknown quantity, and the value of this indicator can be solved by solving the two equations in conjunction. According to the conservation of mass, the transfer of the flow rate, provided that the transfer of the pollutant concentration is stable, has the following relationship.

$$J_{\text{i}} \cdot A_{\text{f}} = E \tag{6.8}$$

where A_{f} is the area of the building floor.

Using the above equation, the gas-phase concentrations corresponding to the pollutants present in the soil around the base of the building can be obtained as follows.

$$C_{\text{isoil}} = C_{\text{isa}} \frac{\frac{D_{\text{i}} A_{\text{f}}}{Q_{\text{soil}} L_{\text{s}}} \left[\text{EXP}\left(\frac{Q_{\text{soil}} L_{\text{f}}}{D_{\text{crack}} A_{\text{crack}}}\right) - 1\right]}{\frac{D_{\text{i}} A_{\text{f}}}{Q_{\text{soil}} L_{\text{s}}} \left[\text{EXP}\left(\frac{Q_{\text{soil}} L_{\text{f}}}{D_{\text{crack}} A_{\text{crack}}}\right) - 1\right] + \text{EXP}\left(\frac{Q_{\text{soil}} L_{\text{f}}}{D_{\text{crack}} A_{\text{crack}}}\right)} \tag{6.9}$$

The total contamination transport E can be obtained by substituting equation (6.9) into Eq. (6.7).

6.2.3 Derivation of Indoor Gas-Phase Concentration and Attenuation Factor

For the indoor space of the building, on the one hand, the concentration of pollutants is constantly diluted due to the continuous invasion of pollutants into the room through vapor, and on the other hand, due to the ventilation of indoor and outdoor. When the indoor concentration reaches a steady-state, there should be the following relationship.

$$C_{\text{ia}} \cdot r_{\text{i}} = E \tag{6.10}$$

where r_{i} is the indoor air exchange rate.

Then the volatile exposure concentration in the room can C_{ia} be obtained. By introducing the building vapor intrusion attenuation factor $\alpha = C_{\text{ia}}/C_{\text{isa}}$ to characterize the overall concentration attenuation from the source to the indoor air, the following equation can be obtained.

$$\alpha = \frac{\frac{D_i A_f}{r_i L_s} \cdot \text{EXP}\left(\frac{Q_{\text{soil}} L_f}{D_{\text{crack}} A_{\text{crack}}}\right)}{\frac{D_i A_f}{Q_{\text{soil}} L_s}\left[\text{EXP}\left(\frac{Q_{\text{soil}} L_f}{D_{\text{crack}} A_{\text{crack}}}\right) - 1\right] + \text{EXP}\left(\frac{Q_{\text{soil}} L_f}{D_{\text{crack}} A_{\text{crack}}}\right) + \frac{D_i A_f}{r_i L_s}} \tag{6.11}$$

In particular, When

$$\frac{Q_{\text{soil}} L_f}{D_{\text{crack}} A_{\text{crack}}} \to 0 \alpha \to \frac{\frac{D_i A_f}{r_i L_s}}{1 + \frac{D_i A_f}{r_i L_s}} \tag{6.12}$$

6.3 Differences Between Volasoil Model and J-E Model

The Volasoil model is a vapor intrusion exposure assessment model developed by the Dutch National Institute for Public Health and the Environment (RIVM) in 2007 to replace the volatilization module of the original CSOIL model. The original CSOIL exposure model was developed to derive intervention values for soil and groundwater clean-up, and in practice, it was found that the volatilization module of the model was not suitable for practical risk assessment. As a result, RIVM has redeveloped the Volasoil model, which is considered the best option between scientific soundness and practical applicability [12] (Fig. 6.3).

6.3.1 Source Concentration Calculation

The three-phase equilibrium in the J-E model is calculated using the Nernst distribution law, i.e., the distribution coefficient method; the three-phase equilibrium in the Volasoil model is calculated using the fugacity model. The mathematical derivation shows that the interconversion of the solid phase (the part adsorbed on soil particles), liquid phase (the part dissolved in soil water), and gas phase (the part volatilized in soil gas) concentrations of pollutants in the subsurface can be achieved. Therefore, although the two models use different principles to calculate the three-phase equilibrium of pollution sources, their essence is the same [13]. However, in practical application, it should be noted that the parameters used in the two standards are not identical because of different calculation formulas. The differences in the values of the parameters may bring about differences in the assessment results. For example, the J-E model uses the Clapton equation to calculate the dimensionless Henry's constant at system temperature, while the Volasoil model calculates the dimensionless Henry's constant at system temperature based on Henry's law. The method used in the Volasoil model usually results in a smaller dimensionless Henry constant than in the J-E model [14].

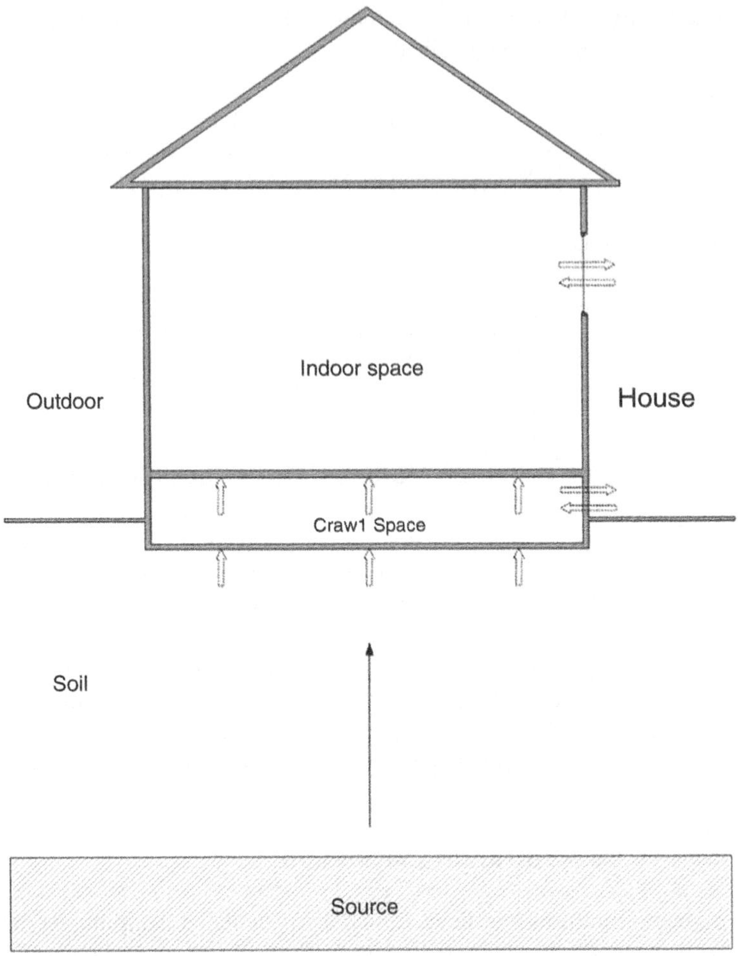

Fig. 6.3 Volasoil model

6.3.2 Calculation of Indoor Air Exposure Concentration

The differences between the J-E model and the Volasoil model for calculating indoor volatile exposure concentrations are mainly in the assumptions about the building structure. The J-E model treats the building as a complete space and considers a part of it to be underground. The Volasoil model considers the building as consisting of two parts, namely the interior part of the building and the duct space at the bottom. Therefore, the J-E model considers the process of pollutant volatilization into the interior of the building in three stages: diffusion from the source to the soil around the bottom of the building, diffusion and convection from the soil around the bottom of the building to the interior, and air exchange with the exterior after entering the interior. The Volasoil model considers the process of pollutant volatilization into the room in four stages: diffusion and convection from the source to the duct space, air exchange with the outdoors after entering the duct space, convection from the duct space to the room, and air exchange with the outdoors after entering the room.

6.4 Study of Vapor Intrusion Risk Assessment

The J-E model and the Volasoil model, as one-dimensional screening models for vapor intrusion risk assessment, contain many simplifying assumptions in the derivation process. The main simplifying assumptions include the following.

1. A single homogeneous soil component or a diffusion coefficient can be simplified to a single homogeneous one.
2. The pollution source concentration is stable, and the whole transfer process reaches a steady-state.
3. The building floor is close to the ground, the house floor covers the area under pressure difference on the existence of only bottom-up gas movement, and the pressure difference is stable.
4. No biodegradable.
5. No advantageous channel.
6. For calculating the gas-phase concentration of the pollution source, the J-E model uses the assumption of three-phase equilibrium based on the distribution coefficient, and the fugacity model used in the Volasoil model is equivalent to it.

These simplifying assumptions above lead to overly conservative model predictions, mainly ignoring the effect of biodegradation, which makes it possible to overestimate the risk by one to three orders of magnitude [15].

Over the past 30 years, a great deal of research work has been conducted around vapor intrusion models. Research has been directed towards improving these 1D screening models above, the development and derivation of 2D models, and the computation of more complex 3D numerical models. While research conducted in the late 1990s and early 2000s focused on fundamental vapor transport processes and attenuation in the subsurface, particularly the role of biodegradation, work in the last decade has focused more on more complex scenarios, such as dominant channels, stratigraphic inhomogeneity, changes in groundwater level and water content, temperature, indoor and outdoor air pressure changes, changes in building characteristics, and safety distances due to biodegradation [16].

Regarding the calculation of gas-phase concentrations of pollutants, some studies consider the binary equilibrium (DED) desorption model closer to the actual situation. This model divides the sorption of organic matter by soil into reversible and irreversible phases and considers that the desorption of the reversible phase follows the principle of linear sorption. In contrast, the desorption of the nonreversible phase can be described by a Langmuir-type isotherm. Nonlinear adsorption would lower the actual desorption process by one to two orders of magnitude than the multiphase equilibrium assumption condition [17].

For the process of pollutant transport in soil, following Johnson and Ettinger's study, Lowell et al. developed a two-dimensional model of the concentration from the source not directly below the building [18], focusing on the effect of horizontal distance on pollutant transport. Abbas et al. described the pollution vapor as the process of concentration depletion over time during forwarding propagation while considering the migration process of the pollutant source itself. The study by Johnson et al. showed that VOCs are significantly influenced by aerobic biodegradation processes in the upward transfer process and diffusion in soil and also by soil particle adsorption [19]. The results of the actual site attenuation evaluation show that the vast majority of site microorganisms are not the limiting factor for natural attenuation capacity, and the key to biodegradation is the assistance of sufficient oxygen in the subsurface to enable aerobic organisms to solidify and abate petroleum hydrocarbon vapors going to shallow soils. The consideration of aerobic biodegradation led to the divergence of the vapor intrusion model into two different lines of study: chlorinated hydrocarbon vapor intrusion (CVI) and petroleum hydrocarbon vapor intrusion (PVI).

For the study of vapor intrusion from chlorinated hydrocarbons, a detailed report was published by the US Environmental Protection Agency (US EPA) in 2012. The report illustrates how different site and building conditions affect vapor intrusion through the simulation results of a three-dimensional numerical model developed by Abreu and Johnson [8, 19]. For chlorinated solvents, source concentration, depth to source, horizontal distance to buildings, conditions of buildings, and subsurface heterogeneity with respect to geological barriers and moisture content are the main influencing factors of interest in the report. The moisture content of the soil is highlighted in this report as a key factor in controlling the upward diffusion of vapor [16]. Other major studies have focused on moisture and soil texture effects on vapor intrusion [2, 20–23]. In addition, several 2D analytical models were developed to predict 2D soil gas concentration profiles, including the vapor distribution around the bottom of the building considering a constant source distribution [2, 23, 24], the effect of source offset [25, 26], the effect of impervious surface cover [27], and lateral transport in stratified soils away from the source edge of the contaminant [28].

The main difference between the risk assessment of petroleum hydrocarbon vapor intrusion and that of chlorinated hydrocarbon vapor intrusion is that petroleum hydrocarbons are biodegradable in the presence of oxygen. This means that the transport of petroleum hydrocarbons is a physical process influenced by biochemical processes and the distribution of oxygen concentrations, which assesses vapor intrusion risk involving petroleum hydrocarbons more complex. A series of one-dimensional steady-state vapor intrusion risk assessment models [29–31] have been developed and refined to support EPA decisions related to petroleum vapor intrusion screening distances. As mentioned above the detailed report published by the USEPA included simulation results from the three-dimensional numerical model developed by Abreu and Johnson [8, 19]. For petroleum hydrocarbon vapors, source concentration, depth to source, horizontal distance to buildings, conditions of buildings, subsurface heterogeneity with respect to geological barriers, and moisture content were the main influencing factors of interest in the report, with particular attention to the role of aerobic biodegradation [16]. Other studies have shown that soil texture and moisture also significantly affect the biodegradation process [32–34], but the associated effects of changes in oxygen distribution due to building size are even more significant. Patterson et al. [35] proposed and demonstrated the following: there is a considerable influence of ground floor building size on vapor intrusion; pollutant concentrations at the edges of buildings are low due to aerobic biodegradation, so pollutant Abreu et al. [36] developed a preliminary two-dimensional model of the variation of vapor intrusion risk with source depth and source concentration for a range of 10 m × 10 m below ground level buildings, taking into account biodegradation. In order to determine an oxygen-lacking zone under a building subslab presence in this scenario, a two-dimensional implicit analytical solution based on instantaneous responses was developed by Knight and Davis [37] and later reduced to an explicit equation to achieve the same result by Verginelli et al. [38]. For the impermeable subslab scenario, as obtained in the results of these works, the EPA suggestion that vertical screening distances would not be applicable to scenarios where large buildings and high source concentrations are involved has been supported. A simple approach was proposed by Yao et al. [39] for the two-dimensional PVI scenario, which coupled an analytical approximation of vapor transport in soil with a piecewise first-order biodegradation model, which is independent of building operating conditions. In addition, an analytical solution for a two-dimensional PVI model was developed by Yao et al. [40] and Verginelli et al. [41], in which diffusion-driven vapor transport in homogeneous soils and piecewise first-order aerobic biodegradation with limitations on oxygen availability were included.

For the second stage of contaminant transport, i.e., the entry of VOCs from the soil around the building into the building, with a three-dimensional numerical model, the presence of preferential channels leads to significant spatial variation in contaminant concentrations in the subslab, even if the subslab has reasonable permeability (e.g., gravel subslab), as shown by Ström et al. [42]. The three-dimensional numerical model applied by Yao et al. investigated the mass flow rate of pollutants into buildings with different foundation slab fracture features [43] and showed that there was no significant role played by the shape and location of the cracks in the foundation or the size of the base slab. Several mathematical relationships were developed by Song et al. [44, 45] to quantify the influence of the effect of stack, wind, and effective leakage area on the rate of entry of soil gas from subslabs and infiltration of outdoor air into residential buildings. A positive correlation between the rate of soil gas entry and the ventilation rate of the building was observed in the simulation results, and the influence of the stack and wind effects on the attenuation coefficient of the subslab was attenuated by this correlation.

As a whole, the difference in VI behavior between hydrocarbon compounds and compounds which are more difficult to degrade (chlorinated hydrocarbons) was clearly demonstrated by modeling studies conducted over the last decade. For hydrocarbon compounds, vapor concentrations could be attenuated by aerobic biodegradation over short distances in the presence of oxygen, and it supports the

US EPA's recommended screening distance criteria for source-to-building. Studies also suggest that for large buildings or where there are high concentrations of hydrocarbons in the source area, these criteria might not be applicable. The attention was focused on the role of subsurface heterogeneity on vapor transport in the case of chlorinated solvents. The moisture content of the soil and fluctuations in the groundwater levels were highlighted as key parameters influencing vapor intrusion into the building above. The effects of wind, the rate of air exchange, and the pressurization of structures on vapor entry into buildings are also explored intensively [16].

6.5 Application of Brownfield Vapor Intrusion Risk Assessment

Based on the J-E model, US EPA and many of its states, as well as ASTM, further introduced a three-phase equilibrium model into the model to calculate indoor air concentrations of VOCs instead of field measurements. However, numerous case studies in the 1990s showed that the results of the J-E model were too conservative after the introduction of the three-phase equilibrium model, so it was only used as a preliminary risk screening, and a "hierarchical-multi-evidence" system was proposed based on the results of soil, groundwater, soil gas, volatile flux, and indoor and outdoor air VOCs monitoring in that order. Multi-lines of evidence assessment technology risk assessment system, and in 2002 [46], issued a technical guideline for health risk assessment of VOCs vapor intrusion exposure (provisional). In 2015, US EPA revised its technical guidelines [47] based on many practical cases in the past 20 years. In particular, it proposed a new method for assessing the risk of vapor intrusion from petroleum hydrocarbon contaminated sites. In particular, a new assessment method was proposed for the risk of VOCs vapor intrusion at petroleum hydrocarbon contaminated sites, using soil gas samples instead of soil sample concentrations to estimate expected indoor air concentrations, taking into account the microbial degradation effect of VOCs in soil gas during migration in the clean unsaturated layer of soil.

Considering biodegradation, US EPA has also proposed distance screening criteria based on monitoring results from numerous cases. In the assessment technical guidelines issued by US EPA in 2002 for chlorinated hydrocarbon VOCs, when the horizontal and vertical distances between the source and the receptor are greater than 100 ft (about 30 m), the exposure pathway is considered nonexistent, and no further investigation and assessment is required. The initial distance screening criteria for petroleum hydrocarbons is 30 ft (about 10 m). 2012 California and 2015 US EPA revised and refined the screening criteria for petroleum hydrocarbon and benzene contaminated sites according to specific conceptual models based on a large number of site measurements and the latest research results, adding 15 ft (about 4.5 m) and 6 ft (about 4.5 m) for different scenarios. The screening distances of 15 ft (about 4.5 m) and 6 ft (about 1.8 m) were added for different scenarios.

On the other hand, given the complexity of the model, beginning in 2015, EPA began using the upper confidence limits of the measured soil gas-indoor air attenuation coefficients and groundwater-indoor air attenuation coefficients as attenuation coefficients for screening levels based on the large amount of actual site data accumulated in the vapor intrusion database and no longer derives them through the J-E model. Based on these statistically determined attenuation factors and toxicological models, EPA developed the Chemical Contaminant Vapor Intrusion Screening Level (VISL) calculator to help agency staff determine soil gas and groundwater vapor intrusion screening levels based on limited initial data. In addition to calculating screening levels, the tool can also calculate indoor air concentrations based on user-entered soil gas and groundwater concentrations, as well as calculate risks from calculated indoor air concentrations and user-supplied indoor air concentrations. Some 2D screening models are also recommended in the relevant technical documents.

It is important to note however that soil gas concentrations collected from open sites will be lower than those collected under buildings due to the capping effect. Unlike risk assessments for built-up sites, future land construction and development of brownfields will result in changes in soil properties and soil gas concentrations; furthermore, there is no way to determine indoor air concentration levels and the ability of buildings to block vapors through actual measurements. This makes brownfield risk assessments more limited in terms of evidence available and more dependent on the use of models.

The Canadian federal government issued the Soil Vapor Monitoring Protocols in 2008 and the Guidance for Soil Vapor Intrusion Assessment at Contaminated Sites in 2010 and drafted a more detailed Guidance Manual for Environmental Site Characterization in Support of Environmental and Human Health Risk Assessment in 2013. In addition to the Canadian government, the Province of Ontario has issued a new Guidance Manual for Environmental Site Characterization in Support of Environmental and Human Health Risk Assessment, which is more detailed (including environmental media such as soil, soil gas, and groundwater). In addition to the federal government, the Province of Ontario has published Draft Technical Guidance: Soil Vapor Intrusion Assessment, similar to the US. The Canadian risk assessment technical approach is based on the experience and technology of the US.

The Generic quantitative risk assessment phase in the UK uses gas screening values as defined in CIRIA C665 and exclusion screening distances for assessing vapor. The UK reviewed ten models, including The J-E Model, The GSI Model, The British Columbia (BC) Model, The Unocal Model, the Modified Johnson Model, the BPRISC Model, The VAPEX3 Model, The Ferguson et al. Model & the Modified Ferguson Model, The VOLASOIL model, and The Jury Model. Most of the models (seven of them) are stationary models incorporating diffusion in the framework of the Johnson and Ettinger (1991) model. The strengths and weaknesses of these models and their applicability to the UK were compared and analyzed, with four of the models available as commercial software (GSI, BC, BPRISC (the J-E model and its sub-models) and the Ferguson models) were subsequently retained for more extensive model validation by comparing measured soil vapor concentrations with predicted concentrations at a number of sites. Ultimately the BPRISC model (incorporating the sub-model of the J-E model) was considered to be the closest to the model characteristics needed to meet their regulatory objectives, are relatively user-friendly and appear to be both mathematically correct and robust. Additionally, in its predictions, the BPRISC model is generally considered to be conservative [48]. Biotransformation sub-models within it should only be used however if there is confidence in the biodegradation process. The model was ultimately used as a computational model to assess the risk of vapor intrusion in the framework of CLEA. Overall, the work on vapor intrusion risk assessments for brownfield redevelopment in the UK also has strong similarities to the US and emphasizes the independence of vapor intrusion assessment, considerable evidence, and a focus on soil gas.

The Netherlands has adopted the Volasoil model as a component of its CSOIl model for vapor intrusion risk assessment. The differences between the Volasoil model and the J-E model are described in Chap. 1. Unlike the above countries, the Netherlands still uses soil/groundwater rather than soil gas to assess vapor intrusion risk. This may be related to the high groundwater table in most cases of the Netherlands, which does not support soil gas sampling.

An important difference from the countries mentioned above is that the soil pollution management process in China has accompanied the development of China's urbanization process, which has resulted in almost all contaminated sites in China being brownfields, with few contaminated sites that are already in a state of utilization publicly available. In the Technical Guidelines for Soil Contamination Risk Assessment of Construction Land (HJ 25.3-2019) promulgated in China, the vapor intrusion process mainly refers to the J-E model under the assumption of three-phase equilibrium and does neither consider biodegradation, nor does it adopt soil gas screening values or screening distances, and only some of the parameters for China Some adjustments were made for China. This has a lot to do with the fact that most contaminated sites in China are brownfields treated by large-scale remediation until the source concentration reaches the standard before further use is allowed and the lack of established sites for further vapor intrusion studies. Since the current management of brownfield sites in China is conservative, it is common for soil remediation units to carry out engineering remediation until the source concentration is reduced to an acceptable level of risk and approved by environmental protection authorities before entering the subsequent land development process, so there are many unknown factors for future land use, resulting in a conservative risk assessment and few cases that can be better integrated with subsequent actual development and utilization scenarios. On the other hand, the typical single use of the J-E model based on soil or groundwater concentration under the assumption of three-phase equilibrium, which lacks the portrayal of the conceptual model, has led to less conservative assessment results of vapor intrusion in some scenarios. Typical examples include (1) VOCs pollutants in the northwest gravel soil layer or southwest rock fissures because gravel and rock have weak adsorption for organic pollutants and the main pollutant mass is assigned in soil gas, which leads to risk underestimation based on soil concentration; (2) high water table areas in the southeastern plains, where deep basements are commonly built in urban development situations, making the water table line is often higher than the basement building floor, which does not meet the scenario assumptions of the J-E model, and the direct infiltration process of contaminated groundwater leads to the underestimation of risk. In fact, some Chinese experts and the Ministry of Ecology and Environment have recognized the limitations of the lack of attention to soil gas and multi-evidence processes, organized several research projects targeted research work, and achieved specific research results, but have not yet been translated into management policies.

6.6 The Future of Vapor Intrusion Risk Assessment in Brownfield Redevelopment

As mentioned earlier, brownfield sites will subsequently undergo redevelopment activities, and their land-use status will change from the status quo, which means that risk assessment of brownfield sites is more difficult to obtain substantial evidence of risk, and it relies more on model predictions; on the other hand, redevelopment and use of land mean that the nature of the soil, building conditions, and sensitive targets on the site will change, and the results of the changes are unknown, which means The uncertainty of risk assessment becomes more significant. This means that further development of assessment prediction models is the most critical research direction for brownfield risk assessment, especially for the most complex vapor intrusion risk assessment.

On the other hand, statistical analysis based on data from a large number of built-up sites can also help predict the risk of brownfield sites after they are built up. However, it is especially important to note that due to social progress, redeveloped brownfield sites tend to have better building conditions and more concrete-oriented building materials than historical built-up sites, which requires more data accumulation on vapor intrusion contamination under current building conditions and some distinction from previous information on wood floor construction.

China, represented by countries undergoing urbanization, is the country most in need of vapor intrusion risk assessment technology in brownfield redevelopment, especially because China has rich soil types and considerable differences in various scenarios and has the most extensive research and application space for related technologies. The problems mentioned above of risk assessment of contaminants between gravels or rock gaps and risk assessment in areas with high water tables are the current difficulties that need to be solved. In addition, due to the short period that China has been concerned about soil contamination, many practitioners, government officials, and the public still lack accurate knowledge of the vapor intrusion process and are either ignorant or over-conservative. This also requires better scientific popularization of vapor intrusion and community risk communication to effectively prevent and control risks while reducing the waste of resources caused by excessive remediation. This requires a better understanding of the science of vapor intrusion and community risk communication to prevent and control the risk and reduce the waste of resources due to over-remediation.

References

1. Rizzo E, Pesce M, Pizzol L, Alexandrescu FM, Giubilato E, Critto A, Marcomini A, Bartke S. Brownfield regeneration in Europe: identifying stakeholder perceptions, concerns, attitudes and information needs. Land Use Policy. 2015;48:437–53.
2. Shen R, Yao Y, Pennell KG, Suuberg EM. Modeling quantification of the influence of soil moisture on subslab vapor concentration. Environ Sci Processes Impacts. 2013;15(7):1444–51.
3. Little JC, Daisey JM, Nazaroff WW. Transport of subsurface contaminants into buildings. Environ Sci Technol. 1992;26:2058–66.
4. Yao Y, Pennell KG, Suuberg EM. Estimation of contaminant subslab concentration in vapor intrusion. J Hazard Mater. 2012;231–232:10–7.
5. Wang X, Unger AJ, Parker BL. Simulating an exclusion zone for vapour intrusion of TCE from groundwater into indoor air. J Contam Hydrol. 2012;140:124–38.
6. Lundegard PD, Johnson PC, Dahlen P. Oxygen transport from the atmosphere to soil gas beneath a slab-on-grade foundation overlying petroleum-impacted soil. Environ Sci Technol. 2008;42(15):5534–40.
7. Millington RJ, Quirk JP. Permeability of porous solids. Trans Faraday Soc. 1961;57(0):1200–7.
8. Abreu LD, Johnson PC. Effect of vapor source-building separation and building construction on soil vapor intrusion as studied with a three-dimensional numerical model. Environ Sci Technol. 2005;39(12):4550–61.
9. Yao Y, Shen R, Pennell KG, Suuberg EM. Comparison of the Johnson– Ettinger vapor intrusion screening model predictions with full three-dimensional model results. Environ Sci Technol. 2011;45(6):2227–35.
10. Bozkurt O, Pennell KG, Suuberg EM. Simulation of the vapor intrusion process for nonhomogeneous soils using a three-dimensional numerical model. Ground Water Monit Remediat. 2009;29(1):92–104.
11. Johnson PC, Ettinger RA. Heuristic model for predicting the intrusion rate of contaminant vapors into buildings. Environ Sci Technol. 1991;25(8):1445–52.
12. Waitz MFW, Freijer JI, Kreule P, Swartjes FA. The VOLASOIL risk assessment model based on CSOIL for soils contaminated with volatile compounds. 715810014. the Netherlands: RIVM; 1996.
13. Wu X, Xie L. Comparative study on Johnson & Ettinger model and Volasoil model in the indoor volatilization risk assessment of contaminant. J Environ Qual. 2012;32(4):984–91.
14. Hulot C, Hazebrouck B, Gay G, Malherbe L, Pokryszka Z. Vapor emissions from contaminated soils into buildings: comparison between predictions from transport model and field measurements. In: International FZK/TNO conference on contaminated soil. 2003. pp. 353–361.
15. Johnson PC, Kemblowski MW, Johnson RL. Assessing the significance of subsurface contaminant vapor migration to enclosed spaces: site-specific alternatives to generic estimates. J Soil Contam. 1999;8(3):389–421.
16. Verginelli I, Yao Y. A review of recent vapor intrusion modeling work. Ground Water Monit Remediat. 2021;41(2):138–44.
17. Zhang R, Jiang L, Zhong M, Han D, Zheng R, Fu Q, Zhou Y, Ma J. Applicability of soil concentration for VOC-contaminated site assessments explored using field data from the Beijing-Tianjin-Hebei urban agglomeration. Environ Sci Technol. 2019;53(2):789–97.
18. Lowell PS, Eklund B(URS Corporation, P. O. B.). VOC emission fluxes as a function of lateral distance from the source. Environ Prog. 2004;23(1):52–8.

19. Abreu LDV, Johnson PC. Simulating the effect of aerobic biodegradation on soil vapor intrusion into buildings: influence of degradation rate, source concentration, and depth. Environ Sci Technol. 2006;40(7):2304–15.
20. Guo Y, Holton C, Luo H, Dahlen P, Johnson PC. Influence of fluctuating groundwater table on volatile organic chemical emission flux at a dissolved chlorinated-solvent plume site. Ground Water Monit Remediat. 2019;39(2):43–52.
21. Qi S, Luo J, O'Connor D, Cao X, Hou D. Influence of groundwater table fluctuation on the non-equilibrium transport of volatile organic contaminants in the vadose zone. J Hydrol. 2020;580:124353.
22. Shen R, Pennell KG, Suuberg EM. A numerical investigation of vapor intrusion — the dynamic response of contaminant vapors to rainfall events. Sci Total Environ. 2012;437:110–20.
23. Yao Y, Mao F, Ma S, Yao Y, Suuberg EM, Tang X. Three-dimensional simulation of land drains as a preferential pathway for vapor intrusion into buildings. J Environ Qual. 2017;46(6): 1424–33.
24. Yao Y, Verginelli I, Suuberg EM. A two-dimensional analytical model of vapor intrusion involving vertical heterogeneity. Water Resour Res. 2017;53(5):4499–513.
25. Shen R, Pennell KG, Suuberg EM. Analytical modeling of the subsurface volatile organic vapor concentration in vapor intrusion. Chemosphere. 2014;95:140–9.
26. Yao Y, Shen R, Pennell KG, Suuberg EM. Estimation of contaminant subslab concentration in vapor intrusion including lateral source–building separation. Vadose Zone J. 2013;12(3): vzj2012.0157.
27. Yao Y, Wu Y, Tang M, Wang Y, Wang J, Suuberg EM, Jiang L, Liu J. Evaluation of site-specific lateral inclusion zone for vapor intrusion based on an analytical approach. J Hazard Mater. 2015;298:221–31.
28. Feng S-J, Zhu Z-W, Chen H-X, Chen Z-L. Two-dimensional analytical solution for VOC vapor migration through layered soil laterally away from the edge of contaminant source. J Contam Hydrol. 2020;233:103664.
29. Verginelli I, Baciocchi R. Modeling of vapor intrusion from hydrocarbon-contaminated sources accounting for aerobic and anaerobic biodegradation. J Contam Hydrol. 2011;126(3–4): 167–80.
30. Verginelli I, Baciocchi R. Vapor intrusion screening model for the evaluation of risk-based vertical exclusion distances at petroleum contaminated sites. Environ Sci Technol. 2014;48(22): 13263–72.
31. Verginelli I, Capobianco O, Baciocchi R. Role of the source to building lateral separation distance in petroleum vapor intrusion. J Contam Hydrol. 2016;189:58–67.
32. Bekele DN, Naidu R, Chadalavada S. Development of a modular vapor intrusion model with variably saturated and non-isothermal vadose zone. Environ Geochem Health. 2018;40(2): 887–902.
33. Luo J, Kurt Z, Hou D, Spain JC. Modeling aerobic biodegradation in the capillary fringe. Environ Sci Technol. 2015;49(3):1501–10.
34. Picone S, Valstar J, van Gaans P, Grotenhuis T, Rijnaarts H. Sensitivity analysis on parameters and processes affecting vapor intrusion risk. Environ Toxicol Chem. 2012;31(5):1042–52.
35. Patterson BM, Davis GB. Quantification of vapor intrusion pathways into a slab-on-ground building under varying environmental conditions. Environ Sci Technol. 2009;43(3):650–6.
36. Abreu LD, Ettinger R, McAlary T. Simulated soil vapor intrusion attenuation factors including biodegradation for petroleum hydrocarbons. Ground Water Monit Remediat. 2009;29(1): 105–17.
37. Knight JH, Davis GB. A conservative vapour intrusion screening model of oxygen-limited hydrocarbon vapour biodegradation accounting for building footprint size. J Contam Hydrol. 2013;155:46–54.
38. Verginelli I, Yao Y, Wang Y, Ma J, Suuberg EM. Estimating the oxygenated zone beneath building foundations for petroleum vapor intrusion assessment. J Hazard Mater. 2016;312:84–96.

39. Yao Y, Yang F, Suuberg EM, Provoost J, Liu W. Estimation of contaminant subslab concentration in petroleum vapor intrusion. J Hazard Mater. 2014;279:336–47.
40. Yao Y, Verginelli I, Suuberg EM. A two-dimensional analytical model of petroleum vapor intrusion. Water Resour Res. 2016;52(2):1528–39.
41. Verginelli I, Yao Y, Suuberg EM. An excel®-based visualization tool of two-dimensional soil gas concentration profiles in petroleum vapor intrusion. Ground Water Monit Remediat. 2016;36(2):94–100.
42. Ström JGV, Guo Y, Yao Y, Suuberg EM. Factors affecting temporal variations in vapor intrusion-induced indoor air contaminant concentrations. Build Environ. 2019;161:106196.
43. Yao Y, Pennell KG, Suuberg EM. Simulating the effect of slab features on vapor intrusion of crack entry. Build Environ. 2013;59:417–25.
44. Song S, Schnorr BA, Ramacciotti FC. Quantifying the influence of stack and wind effects on vapor intrusion. Hum Ecol Risk Assess. 2014;20(5):1345–58.
45. Song S, Schnorr BA, Ramacciotti FC. Accounting for climate variability in vapor intrusion assessments. Hum Ecol Risk Assess. 2018;24(7):1838–51.
46. US EPA. Draft guidance for evaluating the vapor intrusion to indoor air pathway from groundwater and soils (subsurface vapor intrusion guidance). Washington DC: Office of Solid Waste and Emergency Response (OSWER). EPA530-D-02-004.; 2002.
47. US EPA. OSWER technical guide for assessing and mitigating the vapor intrusion pathway from subsurface vapor sources to indoor air. Washington DC: Office of Solid Waste and Emergency Response (OSWER). 9200.2–154.; 2015.
48. Evans D, Herbs I, Wolters RM, Boddington RTB, Hall DH. Vapour transport of soil contaminants. Environment Agency, SP5-018/TR, Research and Development Technical Report. 2002.

The manufacturer's authorised representative in the EU is Springer Nature Customer Service Centre GmbH, Europaplatz 3, 69115 Heidelberg, Germany. If you have any concerns regarding our products, please contact ProductSafety@springernature.com

Printed and bound by CPI Group (UK) Ltd, Croydon, CR0 4YY
30/09/2026
02225259-0006